STEAM AROUND CARLISLE IN THE 1960S

KEITH W. PLATT

AMBERLEY

First published 2024

Amberley Publishing
The Hill, Stroud
Gloucestershire, GL5 4EP

www.amberley-books.com

ISBN 978 1 3981 0564 5 (print)
ISBN 978 1 3981 0565 2 (ebook)

British Library Cataloguing in Publication Data.
A catalogue record for this book is available from the British Library.

Origination by Amberley Publishing.
Printed in the UK.

Introduction

In the late 1950s and early 1960s, my youthful enthusiasm for railways in general and steam locomotives in particular was at first concentrated around waiting by the local lineside with a group of like-minded friends collecting engine numbers. We were restricted in where we were able to go because of our age and the amount of pocket money we could muster. Bicycles gave us a certain amount of freedom to be able to journey further afield to see shiny Pacifics hauling express passenger trains, rather than the endless procession of heavy coal and steel trains. The lure of seeing different classes of engines from different regions grew stronger with each train-spotting trip. As pocket money and the ability to earn it became more readily available, so the boundaries of our possible destinations increased. Planning our journeys, studying railway timetables and checking allocations of engines at different sheds all became essential parts of preparing for the trip.

It now seems amazing that a group of ten-year-olds could safely travel from our homes north of Doncaster and spend a Saturday roaming around the stations and engine sheds in places such as Sheffield, Leeds, Derby and York. As ten-year-olds grew to young teenagers, so our horizons broadened – Birmingham and Wolverhampton to see the final days of the 'Kings', Tamworth Junction and Lichfield, the closest locations for us to see the delights of the West Coast Main Line. All these adventures depended on how far we could go and return in a day by train, which was the final barrier to overcome.

As fifteen-year-olds, our group had dwindled to myself and one friend. The others had found new interests and activities as well as the world of work to occupy their time. A borrowed college scarf became the key to a whole new set of destinations as railway timetables were replaced by Esso road maps. By simply waving our thumbs at traffic at the end of the road we were able to end up anywhere we chose to go. In our case, one of the first and repeat destinations was Carlisle.

Carlisle was a place we had visited by rail in the past but it involved three changes of train and left us with just under three hours in the city. However, it gave us a tantalising taste of what was on offer at this important and busy junction serving so many different lines.

Midnight was the preferred departure time for our hitch-hiking expeditions to Carlisle because there was considerable heavy overnight goods traffic both on the A1 and the A66. We could usually get to Carlisle by the time the cafes had opened for breakfast. In the summer of 1964, the last year of the 'Coronation' Pacifics, the two of us spent a week on the lineside of the West Coast Main Line. With the use of sleeping bags and a small tent we worked our way up the line from Preston to Carlisle visiting depots on the way and finding suitable locations for overnight stays.

Eventually it came time for my friend to exchange his borrowed college scarf for a real one. By then we had met and befriended three older local enthusiasts, who travelled by car to watch and photograph steam. I became a regular companion on their journeys to see out the last days of steam.

I have tried to select images which portray the railway scene around Carlisle and have included locations on railway lines which had a direct connection to Carlisle and its motive power allocation.

In the final years of steam, the variety of different classes of locomotives active around Carlisle reduced dramatically, and certain classes were concentrated at the city's two remaining depots. In January 1967, after Carlisle Upperby had closed to steam, Carlisle Kingmoor still had 129 steam locomotives on its books. Of these, thirty were 'Britannia' Pacifics and twenty-four were 9F 2-10-0s, but by December 1967 the depot had closed and steam had been all but eliminated from the area.

All the images in the book are original slides and negatives from my collection with the names of the photographers where this is known.

I would like to thank my wife, Andrea, for her proofreading work, IT expertise, support and advice.

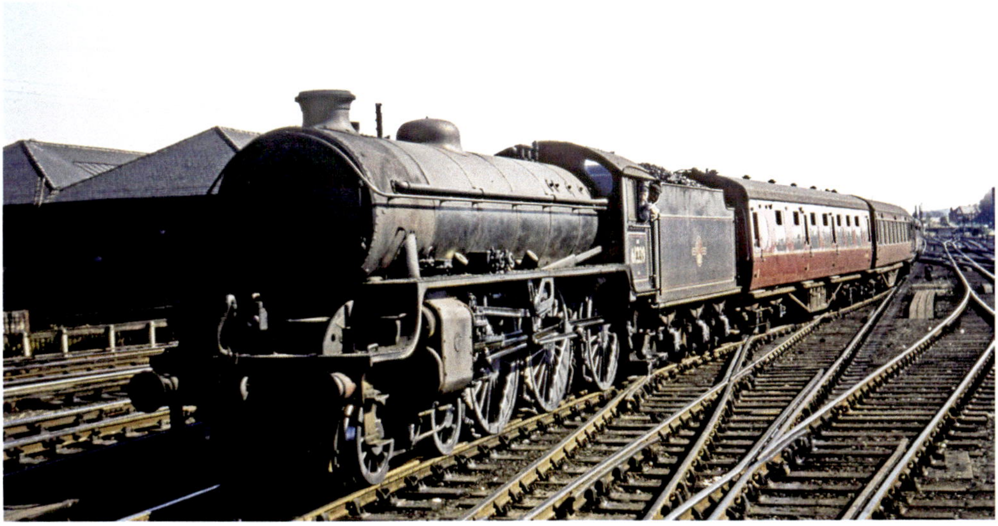

5 March 1960 LNER B1 4-6-0 61239 heads into Carlisle from the south.
LNER B1 4-6-0 61239 was built by the North British Locomotive Company of Glasgow for the LNER in October 1947. It spent its first few years in service working from York and Neville Hill depots before being transferred to Carlisle Canal in September 1953. After this transfer it became the responsibility of Cowlairs Works for its overhauls and it quickly acquired the usual Scottish B1 adornment of curved underside corners to the running plate. The locomotive remained at Carlisle Canal depot until it moved to Gorton MPD in June 1962 but was withdrawn less than two months later in August 1962.

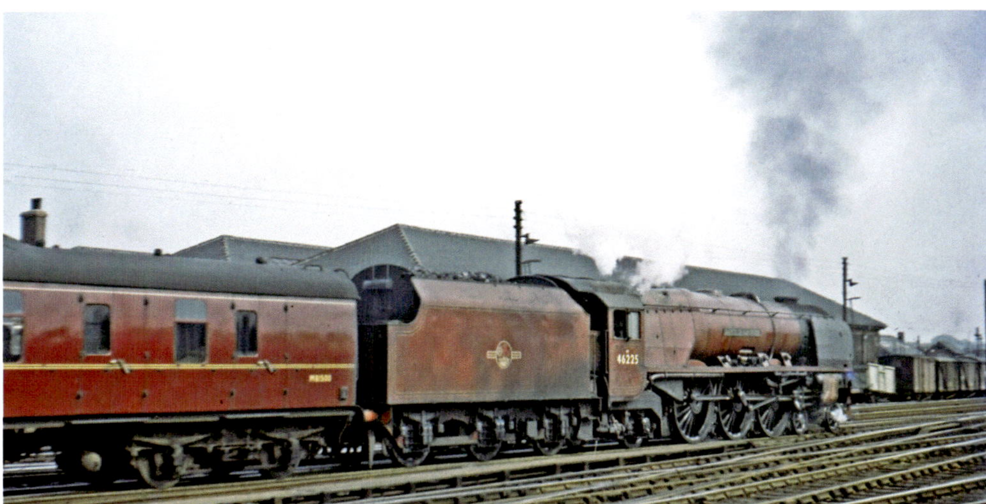

May 1960 LMS Coronation 4-6-2 46225 departs south from Carlisle Citadel.
LMS Coronation 4-6-2 46225 *Duchess of Gloucester* had been built at Crewe Works and entered service in May 1938. It had spent most of its career based at either Camden or Crewe North depots before its reallocation to Carlisle Upperby MPD in June 1959. It remained active from this depot until withdrawal in September 1964, the same year it received a diagonal yellow warning stripe barring it from working south of Crewe. By the end of the year, it had been removed to Arnott Young at Troon for scrapping in their yard.

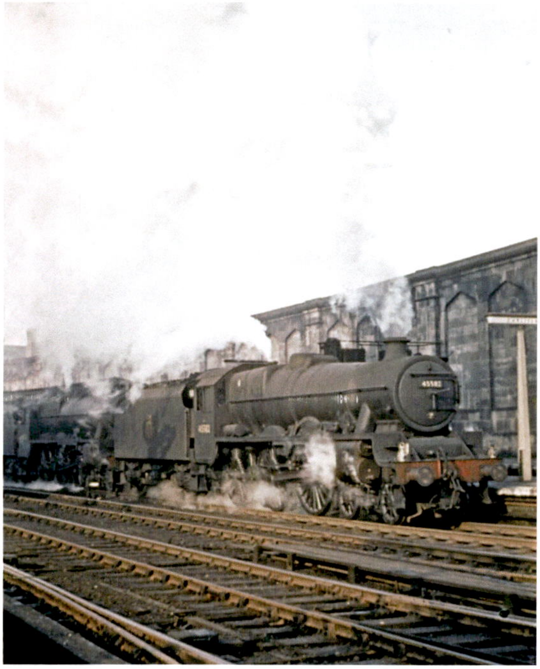

June 1960 LMS Jubilee 4-6-0 45582 at Carlisle Citadel.
LMS Jubilee 4-6-0 45582 *Central Provinces* at Carlisle Citadel was built by the North British Locomotive Company of Glasgow for the LMS in November 1934. At the time of the photograph, it was a Preston-based locomotive and was seen double-heading with a classmate away from Carlisle Citadel. In September 1961 it was transferred to Carnforth but was withdrawn in December 1962 and scrapped at Crewe Works in the summer of 1963.

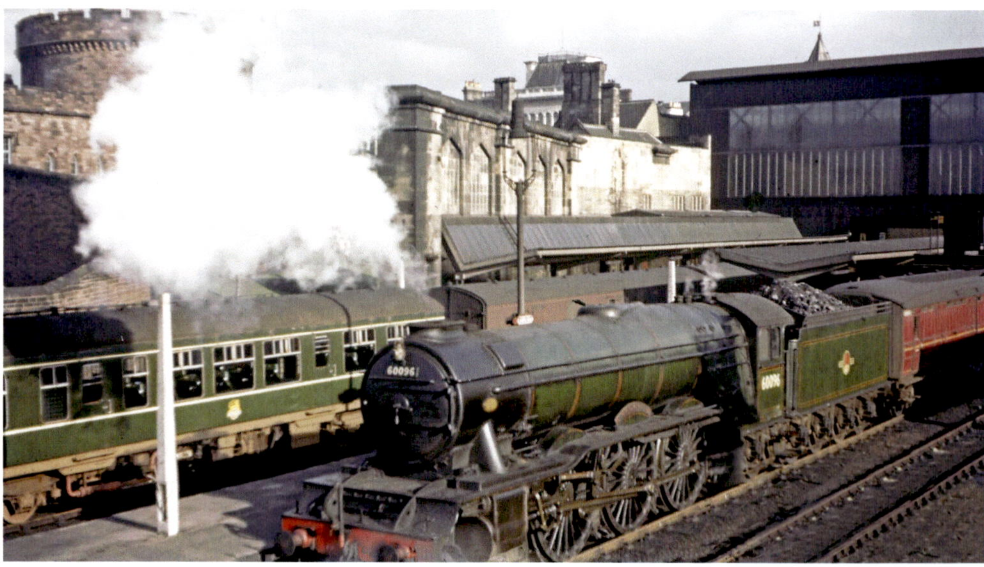

June 1960 LNER A3 4-6-2 60096 heads north from Carlisle.
LNER A3 4-6-2 60096 *Papyrus* was built at Doncaster Plant in February 1929 and was allocated to several depots along the East Coast Main Line in its thirty-four-and-a-half years of active service. In July 1950 it became a resident of Haymarket MPD and remained there until its final move in December 1961 took it to St Margaret's depot. It received a double chimney as part of an overhaul in May 1958 and during its final overhaul, at its birthplace, it was fitted with trough-type smoke deflectors. It was finally withdrawn in September 1963 and spent almost a year in store, first at Cowlairs Works and later at Parkend MPD. It was scrapped in July 1964 by Arnott Young of Carmyle.

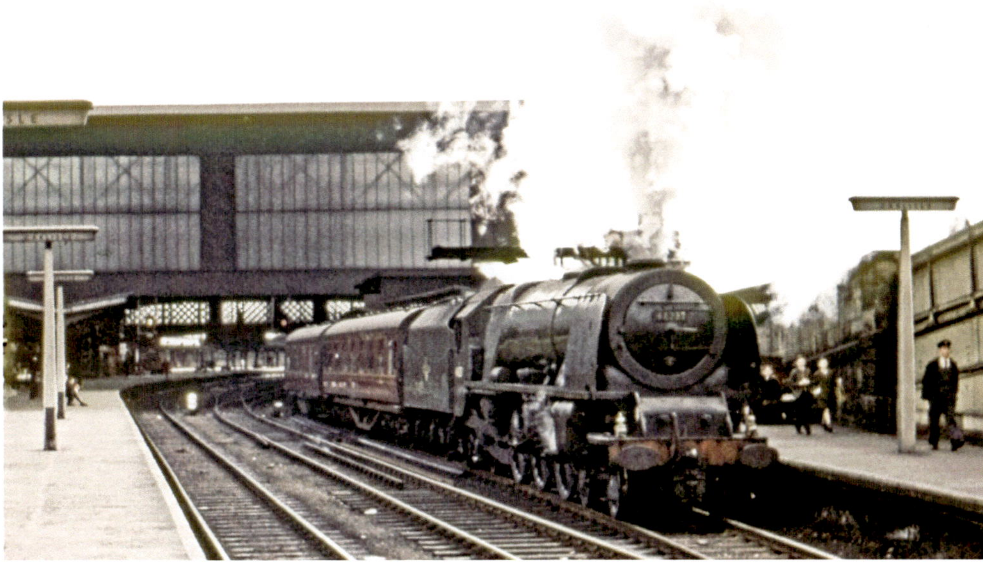

April 1961 LMS Coronation 4-6-2 46237 departs Carlisle Citadel.
LMS Coronation 4-6-2 46237 *City of Bristol* was built at Crewe Works in August 1939 and during its early career was based at both Camden and Crewe. It was transferred north to Carlisle in June 1958 to Upperby MPD, but spent a year working from Kingmoor depot before it returned to Upperby's allocation in April 1962. Along with several remaining Coronation Pacifics it was withdrawn in September 1964 and was cut up by Arnott Young of Troon by the end of the year.

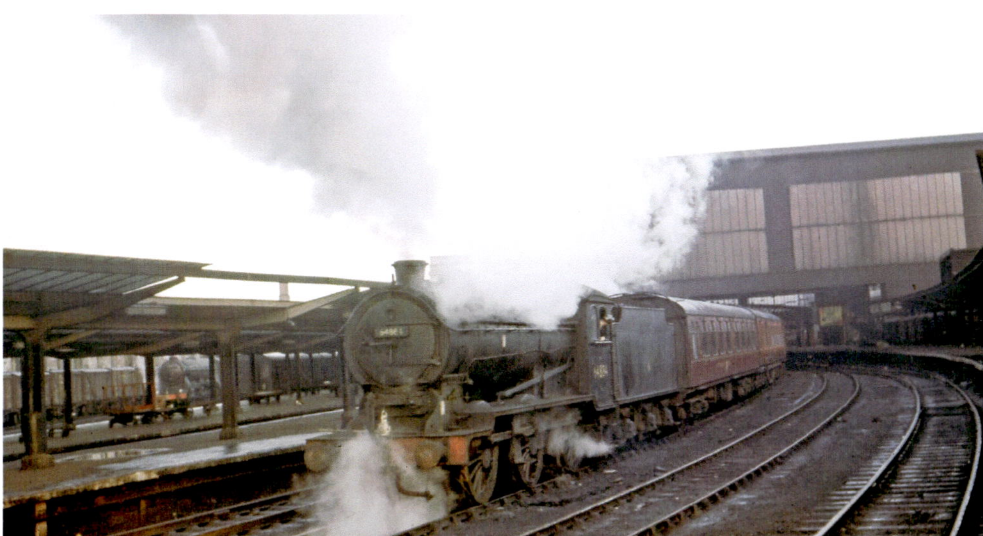

April 1961 LNER J39 0-6-0 64884 shunts stock at Carlisle Citadel.
LNER J39 0-6-0 64884 had been built at Darlington Works in August 1935. It spent the first seven years of its service at Doncaster MPD before it moved to the Great Eastern. In September 1947 it was transferred to Carlisle Canal and began operating on the LNER routes out of Carlisle. It was withdrawn from service in March 1962 and was taken to Cowlairs Works where it was scrapped in the summer of 1962.

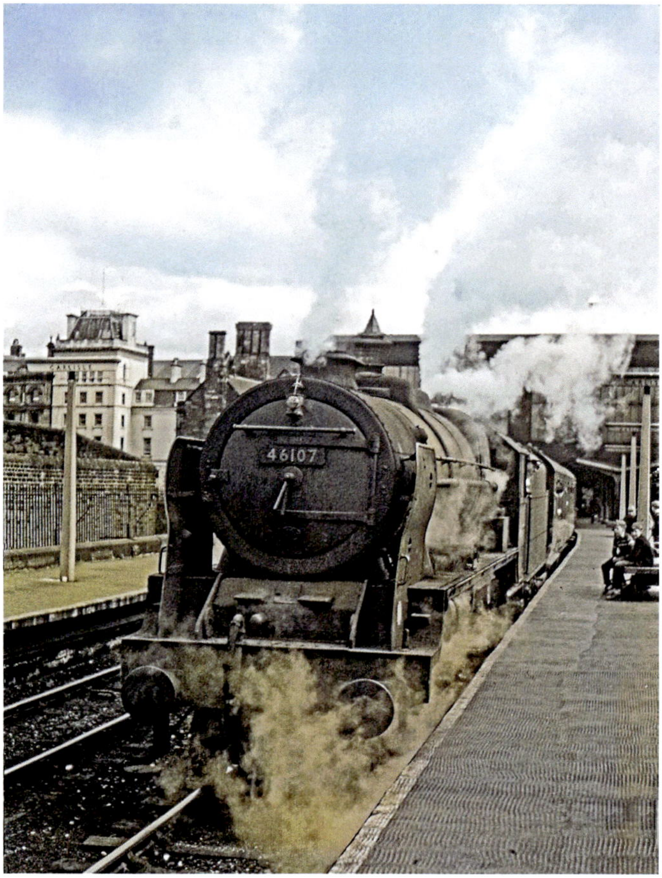

20 May 1961 LMS Royal Scot 4-6-0 46107 departs north from Carlisle Citadel. LMS Royal Scot 4-6-0 46107 *Argyll & Sutherland Highlander* had been built in September 1927 by the North British Locomotive Company of Glasgow for the LMS. It had spent a lot of its early career based in Scotland and at the time of nationalisation it was on the books of Polmadie MPD. It remained there until withdrawal in December 1962. As with all other members of the Royal Scot class, its appearance was very much altered by the fitting of a tapered boiler and double chimney in 1950. After withdrawal it spent quite a long time in store at Parkhead MPD and was finally scrapped by McWilliam of Shettleston in August 1964.

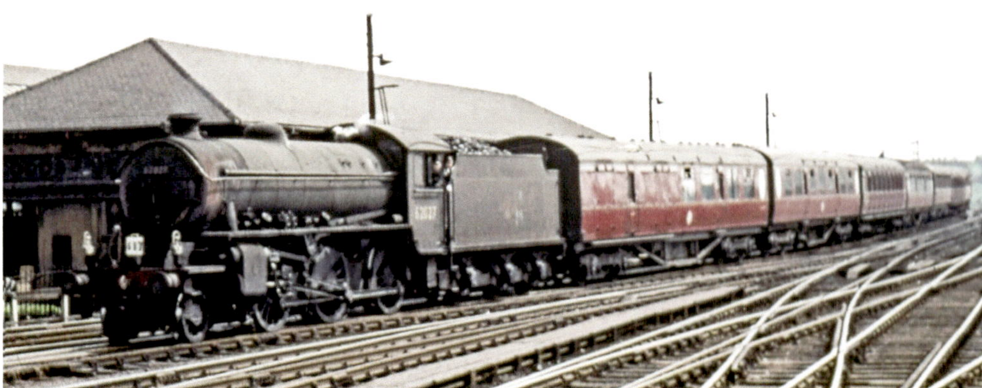

20 May 1961 LNER K1 2-6-0 62027 draws into Carlisle Citadel.
LNER K1 2-6-0 62027 had been built in August 1949 by the North British Locomotive Company of Glasgow for British Railways to the design of the LNER. It had been allocated to Blaydon MPD from new until May 1962 when it moved to Consett depot. One final transfer took place in May 1965 to Blyth North from where the locomotive was withdrawn in March 1967. It was cut up by Arnott Young of Dinsdale in the spring of 1967. It was photographed as it headed into Carlisle Citadel passing the former LNWR Crown Street goods depot.

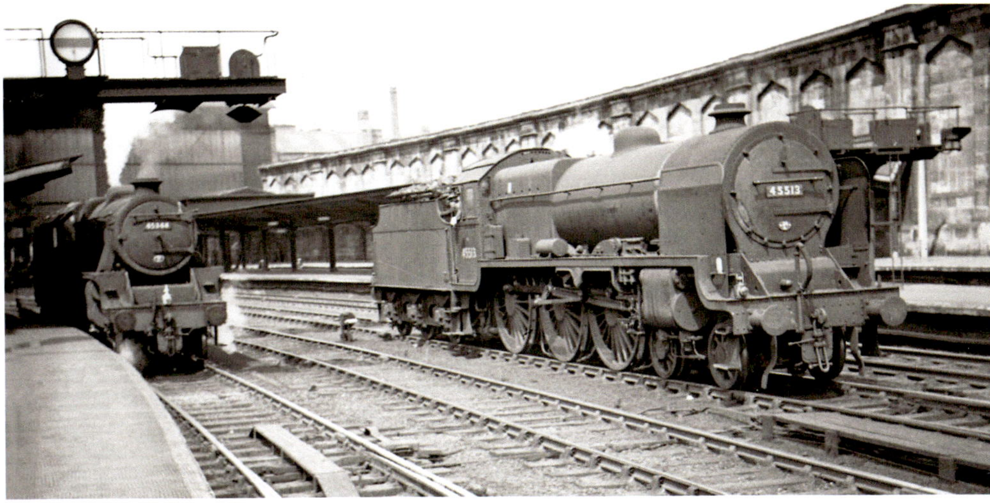

June 1961 LMS Patriot 45513 and LMS Black Five 45344 at Carlisle Citadel.
LMS Patriot 4-6-0 45513 was one of the un-rebuilt members of the class and had emerged from Crewe Works in September 1932. It had spent most of its career based at Crewe and at various locations in the North West. In January 1955 it was transferred to Carlisle Upperby but was then moved again to Carnforth in September 1960. Its final reallocation was to Edge Hill MPD in September 1961 and it was withdrawn from there in September 1962. LMS Black Five 4-6-0 45344 reverses back to attach to its train. It was a Carlisle Upperby-based locomotive at the time but was to move on to Carnforth later in the month. It ended its working life at Croes Newydd MPD in August 1966 and was cut up by Cashmore's of Great Bridge in March 1967.

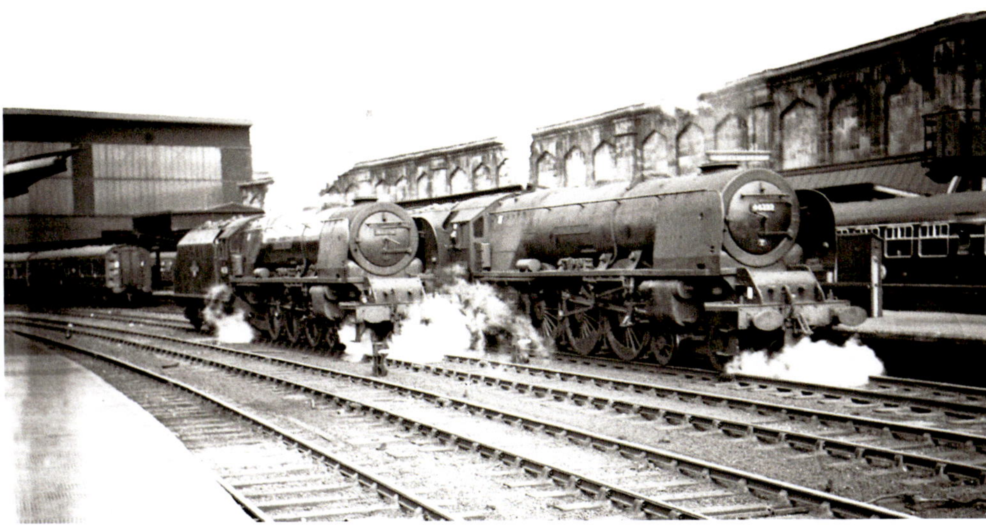

June 1961 LMS Coronation 4-6-2s 46222 and 46236 at Carlisle Citadel.
LMS Coronation 4-6-2 46222 *Queen Mary* comes off its southbound express to be replaced by 46236 *City of Bradford*. 46222 had been a Polmadie-based locomotive throughout its British Railway's career. It had been built at Crewe Works in June 1937 and was withdrawn in October 1963. 46236 was another Crewe product dating from July 1939 and had spent much of its career based at Camden and Crewe. In March 1961 it was reallocated to Carlisle Upperby and remained active until withdrawal in March 1964.

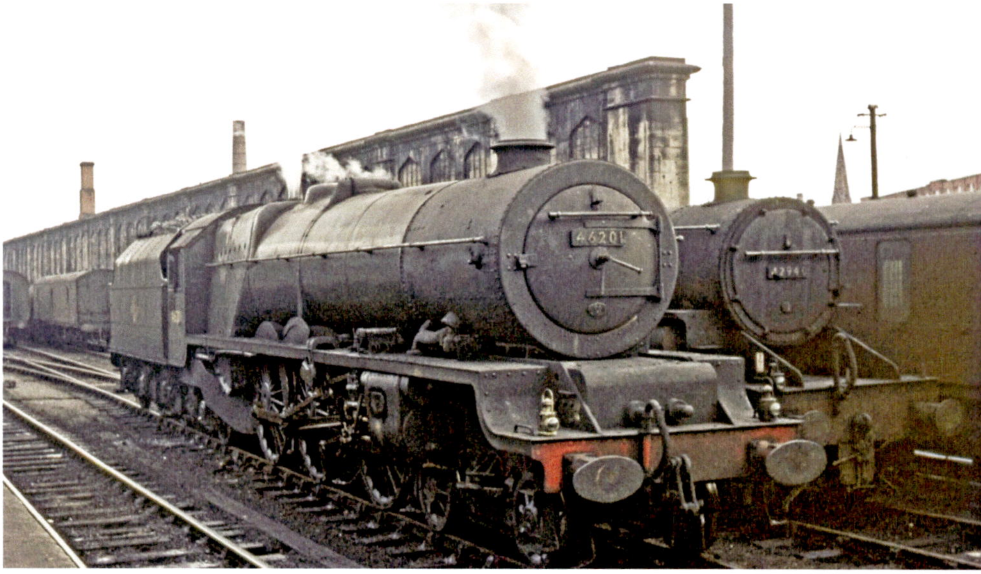

8 September 1961 LMS Princess Royal 4-6-2 46201 with LMS Crab 2-6-0 42940 at Carlisle Citadel. LMS Princess Royal 4-6-2 46201 *Princess Elizabeth* had been built at Crewe Works in November 1933 and had been allocated to several major depots on the London Midland and Scottish Railway. In March 1961 it was reallocated to Carlisle, first to Kingmoor and then in January 1962 to Upperby depot. It was withdrawn in October 1962 and eventually purchased privately for the Princess Elizabeth Locomotive Society. LMS Crab 2-6-0 42940 was another product of Crewe Works and was released into traffic in December 1932. It was a Crewe South-allocated locomotive at the time of the photograph and was withdrawn in September 1965 from Stockport Edgeley. It was cut up by T. W. Ward of Killamarsh to help feed the Sheffield blast furnaces situated close by.

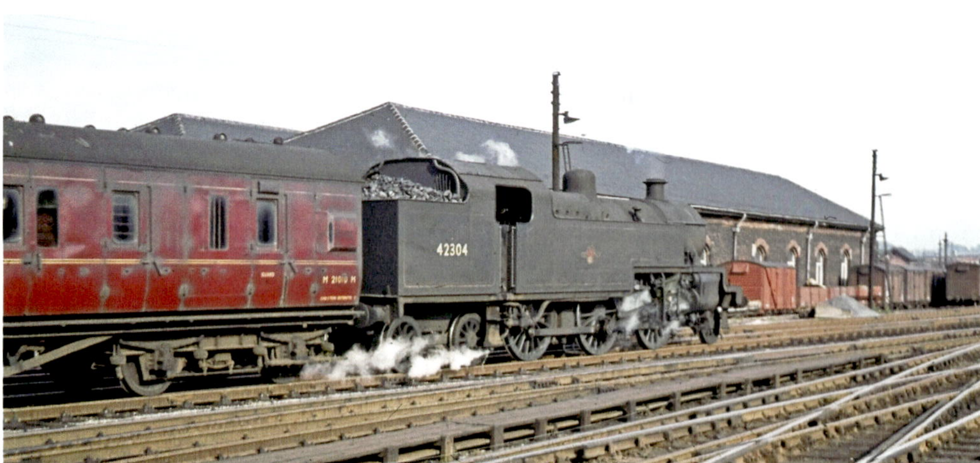

12 July 1962 LMS Fowler 4MT 2-6-4T 42304 departs south from Carlisle Citadel. LMS Fowler 4MT 2-6-4T 42304 had been built at Derby Works in January 1928 and spent its early career based at depots in London and the North West. In March 1960 it was transferred to Carlisle Kingmoor and was withdrawn from there in September 1962. It was photographed as it departed Carlisle Citadel passing the former LNWR Crown Street goods depot.

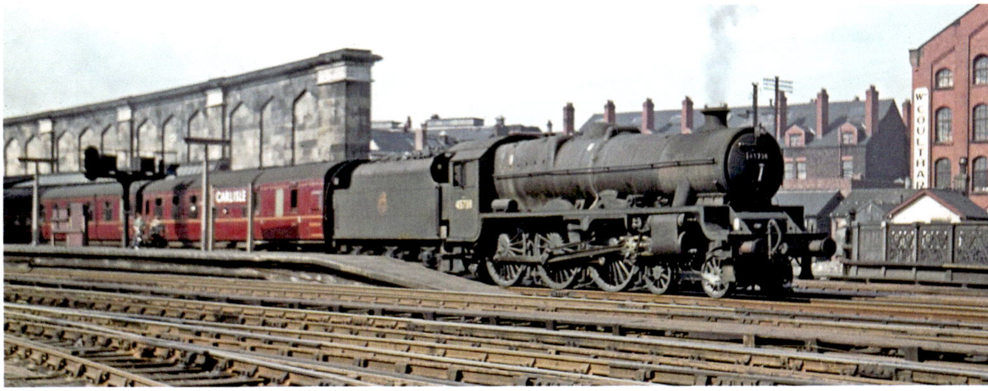

12 July 1962 LMS Jubilee 4-6-0 45738 departs from Carlisle Citadel.
LMS 'Jubilee' 4-6-0 45738 *Samson* had been built at Crewe in December 1936 and spent time at both Crewe North and Bushbury during its active life. In November 1959 it was transferred to Carlisle Upperby and in July 1962 moved across to Kingmoor depot from where it was withdrawn in December 1963. It was photographed as it departed Carlisle Citadel within a few months of it being taken out of service. Of note is the fact that it still carried the 'cycling lion' emblem on its tender which had been superseded from 1956 as locomotives went through overhaul.

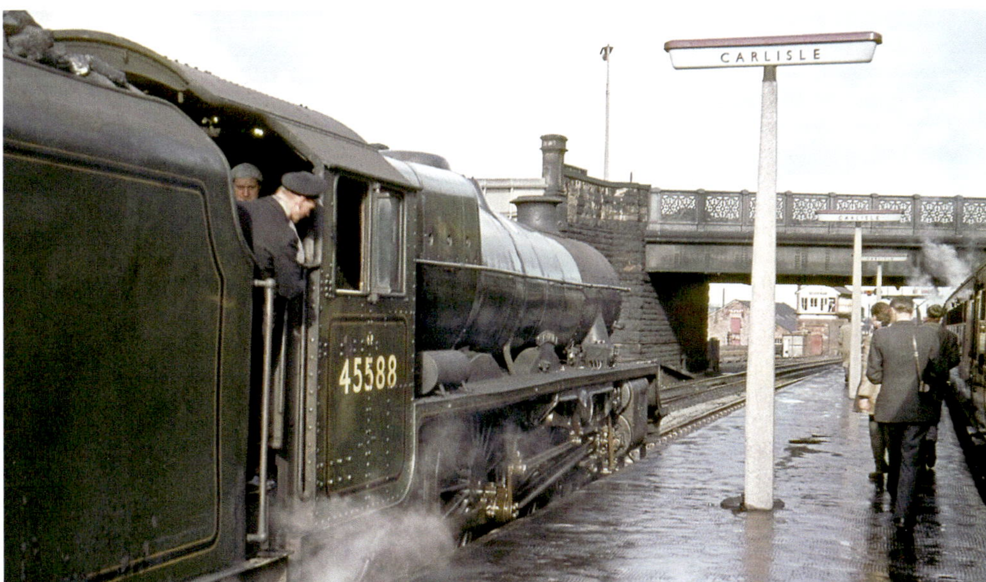

15 April 1963 LMS Jubilee 4-6-0 45588 at Carlisle Citadel.
LMS Jubilee 4-6-0 45588 *Kashmir* had been built by the North British Locomotive Company of Glasgow for the LMS in December 1934 and had been a Carlisle-based locomotive since November 1957. At the time it was a Carlisle Upperby engine and moved to Kingmoor in July 1962 until withdrawal in May 1965. 45588 was seen at Carlisle Citadel at the head of the Stephenson Locomotive Society's *Scottish Rambler* which it hauled from Carlisle to Dumfries. The tour was a massive four-day event departing from Glasgow and travelling over many branch lines. It involved the use of fourteen different steam locomotives and encountered a blizzard on the first day.

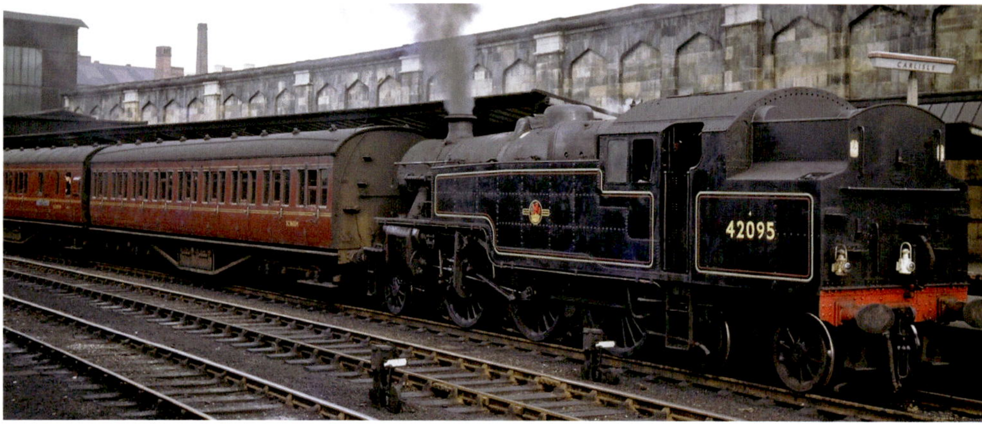

April 1963 LMS Fairburn 4MT 2-6-4T 42095 arrives at Carlisle Citadel.
LMS Fairburn 4MT 2-6-4T 42095 had been built at Brighton in June 1951, three-and-a-half years after the formation of British Railways. For the first eight years of its service it was based at various Southern Region depots and then on to Watford before its reallocation to Carlisle Canal in July 1961. Carlisle Canal MPD was responsible for supplying and servicing the motive power for service on the ex-LNER routes out of Carlisle. The depot closed in June 1963 and these duties passed to Carlisle Kingmoor. 42095 was reallocated in May 1963 to take up banking duties from Tebay MPD and was withdrawn from there in June 1966.

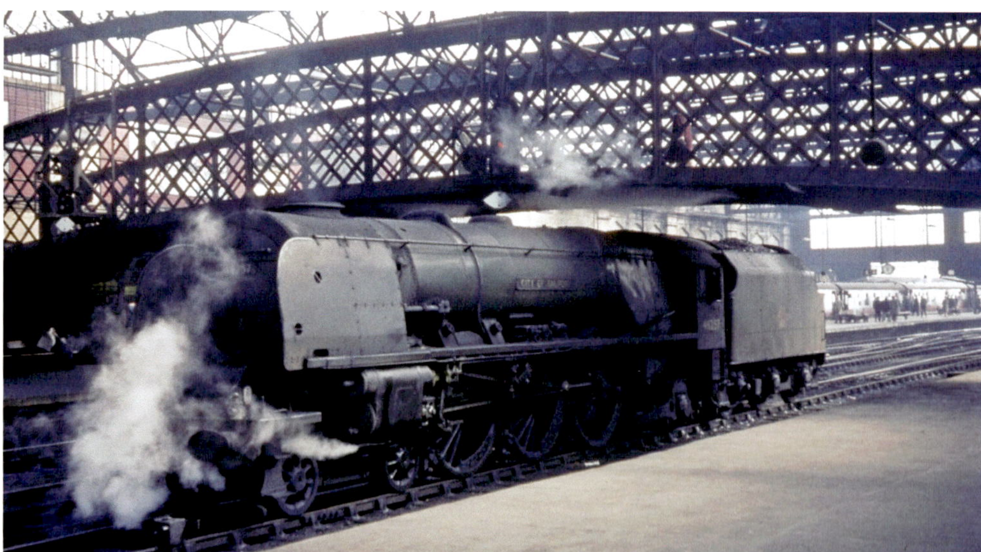

May 1963 LMS Coronation 4-6-2 46257 at Carlisle Citadel.
LMS Coronation 4-6-2 46257 *City of Salford* was built by British Railways at Crewe Works in May 1948. It was the last of the class and incorporated the modification by H. G. Ivatt to Stanier's original design. It was allocated to Camden from new and after ten years based here moved to Carlisle Upperby in October 1958. One final reallocation saw it move across the city to Kingmoor in April 1961. It was withdrawn along with the other survivors of the class in September 1964. It has been rumoured that when it arrived at Arnott Young of Troon for scrapping in January 1965, such was its immaculate condition that the workers got in touch with their manager to ensure that it was actually for cutting up.

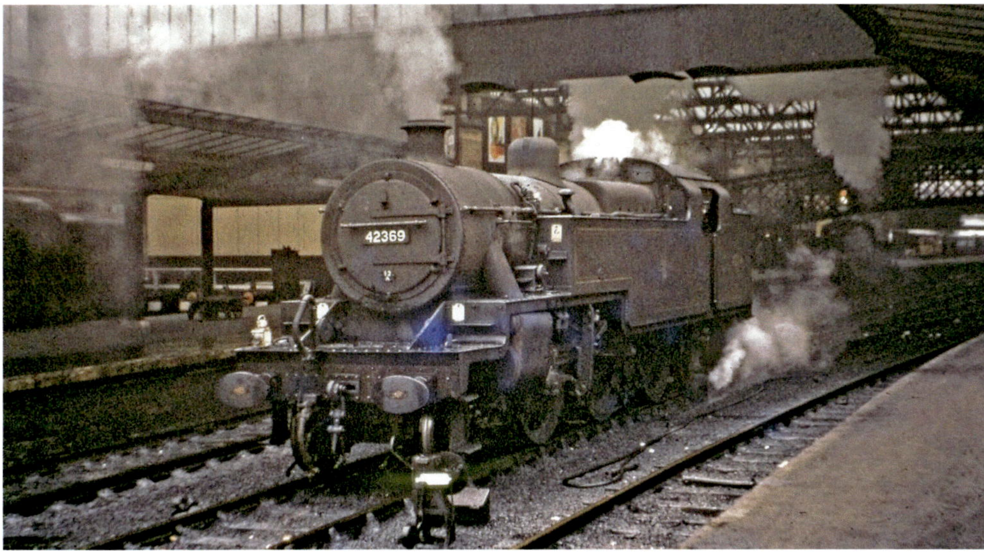

May 1963 LMS Fowler 4MT 2-6-4T 42369 at Carlisle Citadel.
LMS Fowler 4MT 2-6-4T 42369 had been built at Derby Works in September 1929 and had spent much of its working life at depots in the North West. In June 1962 it was reallocated to Carlisle Upperby and four months later across to Kingmoor depot. In March 1964 it left Carlisle for Southport MPD and was eventually withdrawn from Springs Branch, Wigan, in May 1965. It was cut up by Central Wagon of Ince in October of that year. (Photograph L. Flint)

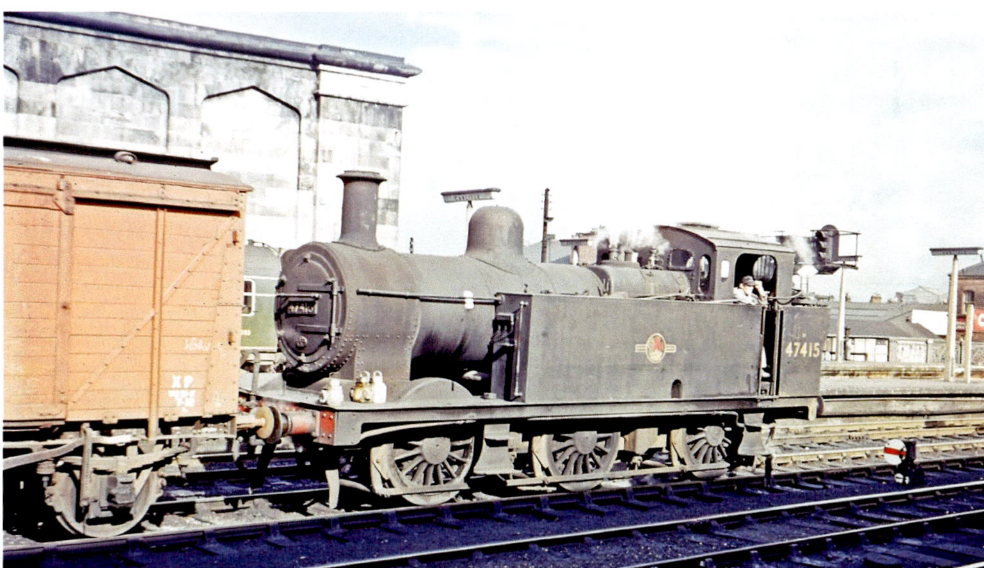

18 August 1963 LMS Jinty 3F 0-6-0T 47415 shunts at Carlisle Citadel.
LMS Jinty 3F 0-6-0T 47415 had been built by the Vulcan Foundry for the LMS in December 1926. Throughout its career with British Railways, it had been allocated to Carlisle Upperby MPD and engaged in shunting movements around Carlisle Citadel and the nearby marshalling yards. In April 1964 it was transferred to Edge Hill, Liverpool, and was withdrawn from there in August 1966.

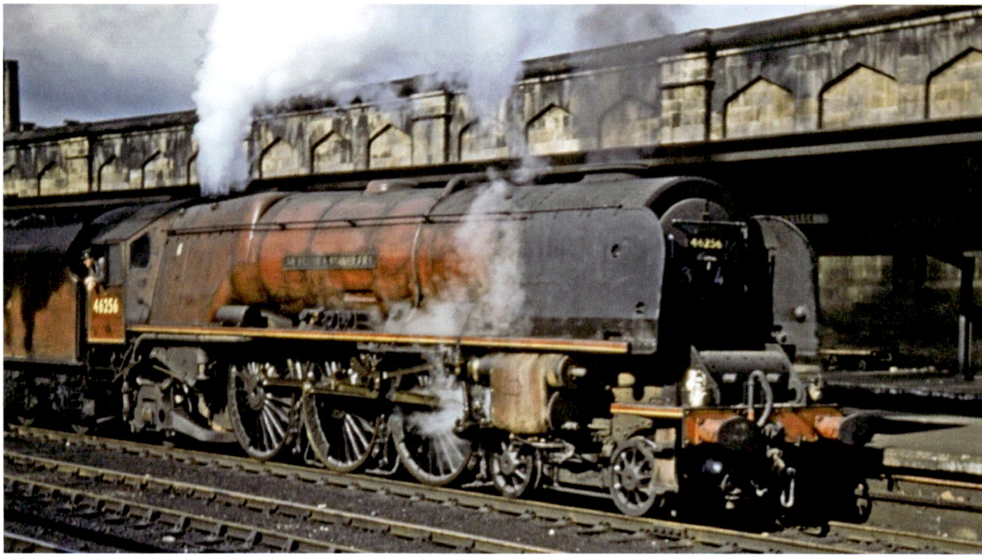

September 1963 LMS Coronation 4-6-2 46256 departs Carlisle Citadel.
LMS Coronation 4-6-2 46256 *Sir William A. Stanier FRS* had been built at Crewe Works and emerged just before nationalisation in December 1947. It had spent its entire career based at either Camden or Crewe North depots except for a four-month spell at Carlisle Upperby from November 1959. It was finally withdrawn from Crewe in October 1964, a few weeks after the other survivors of the class, in order to work a farewell special. It was cut up by Cashmore's of Great Bridge in March 1965. It was seen as it departed Carlisle Citadel on an express.

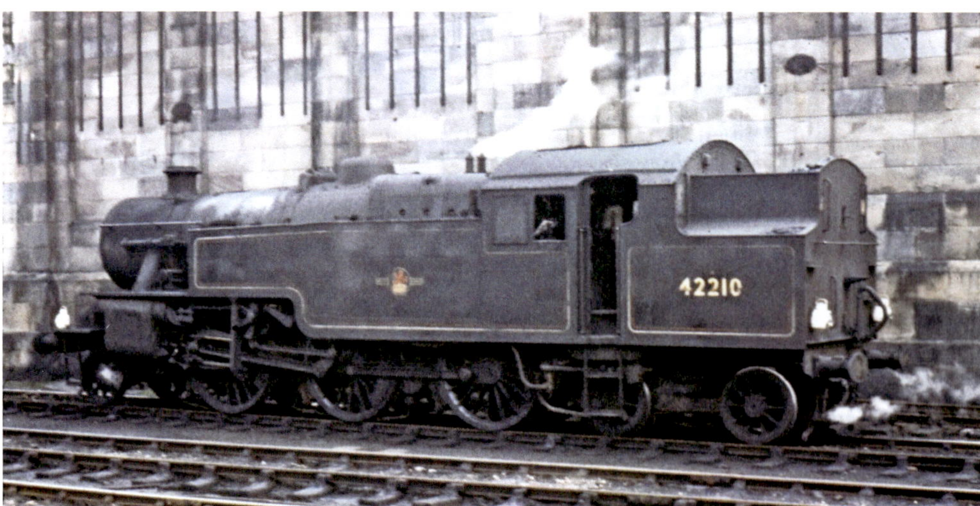

18 April 1964 LMS Fairburn 4MT 2-6-4T 42210 waits at Carlisle Citadel.
LMS Fairburn 4MT 2-6-4T 42210 was built at Derby Works in December 1945 and was reallocated to Carlisle Canal depot as one of the replacements for older, ex-LNER locomotives which had been withdrawn from service. With the closure of the depot in June 1963 the engine was transferred across to Kingmoor MPD and within a year had moved again to Tebay depot. It was withdrawn from Tebay in May 1967 and scrapped by J. McWilliam of Shettleston in early 1968.

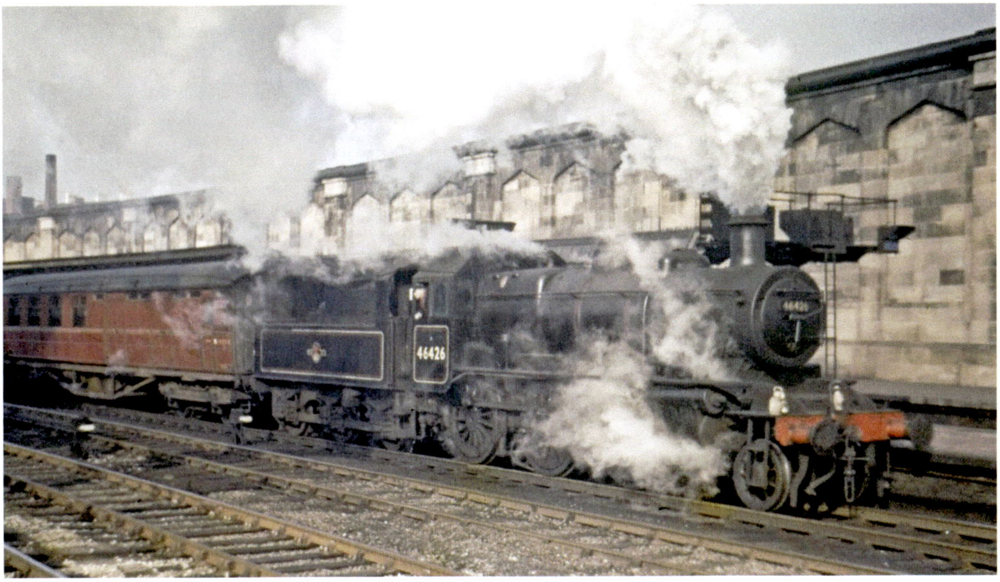

May 1964 LMS Ivatt 2MT 2-6-0 46426 moves coaching stock at Carlisle Citadel.
LMS Ivatt 2MT 2-6-0 46426 had been built for British Railways by Crewe Works in December 1948. It was allocated first of all to Rhyl MPD and was transferred to Lancaster Green Ayre in October 1956. Its final move came in January 1964 which saw it become a Carlisle Upperby loco and was withdrawn from there in September 1966.

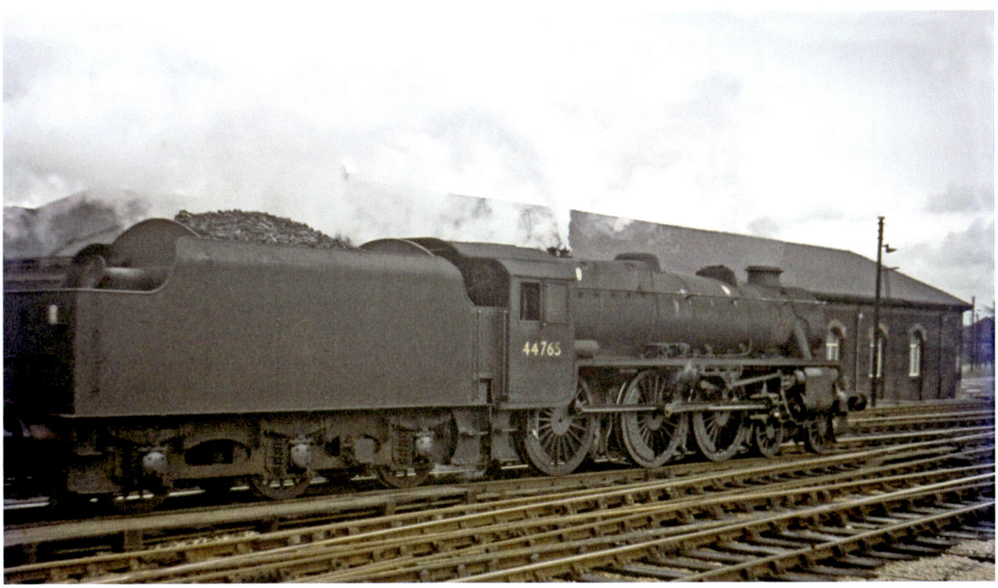

June 1964 LMS Black Five 4-6-0 44765 departs from Carlisle.
LMS Black Five 4-6-0 44765 had been built at Crewe in December 1947 and was one of only six Black Fives to be fitted with a double chimney. It spent its entire working life based in Crewe at Crewe North until May 1965 when it was transferred to Crewe South depot. It was finally withdrawn from there in September 1967 and cut up by Cashmore's of Great Bridge in March 1968.

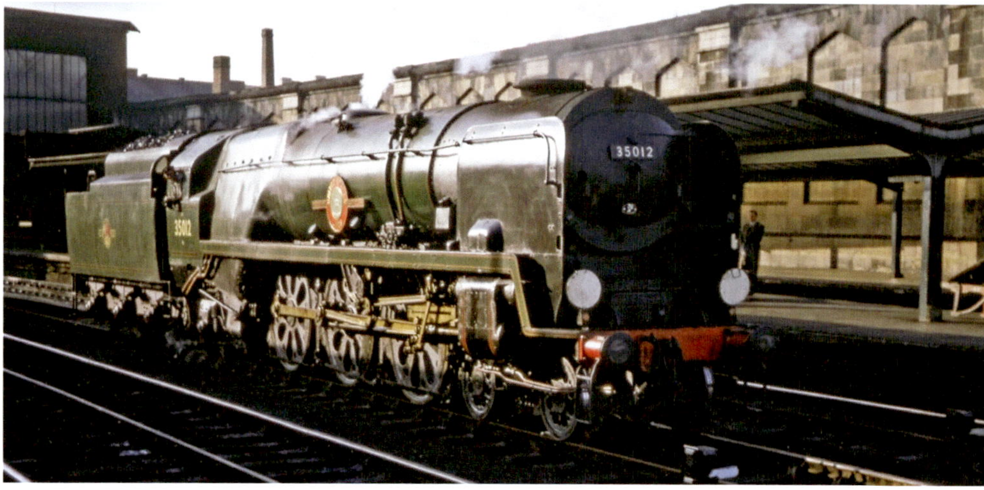

13 June 1964 SR Merchant Navy 4-6-2 35012 at Carlisle Citadel.
SR Merchant Navy 4-6-2 35012 *United States Lines* had been built at Eastleigh Works in January 1945 and rebuilt there in its final form in 1957. It had been a Nine Elms-allocated locomotive for most of its career apart from a few years spent at Weymouth. It was at Carlisle Citadel awaiting the arrival of the 'Solway Ranger' which had been organised by the West Riding branch of the RCTS. The train was hauled by 35012 from Leeds to Penrith via Carnforth and then hauled the tour back to Leeds from Carlisle. (Photograph L. Flint)

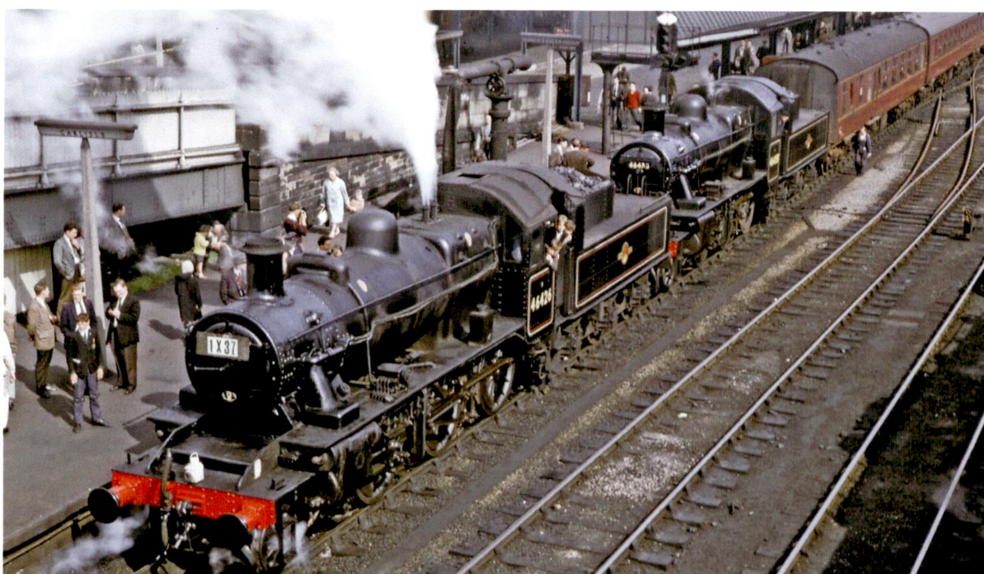

13 June 1964 LMS Ivatt 2MT 2-6-0s 46426 and 46458 on the 'Solway Ranger' at Carlisle.
LMS Ivatt 2MT 2-6-0s 46426 and 46458 had arrived at Carlisle Citadel at the head of the Solway Ranger Rail Tour empty stock. The tour had been hauled from Leeds City to Penrith by Merchant Navy 4-6-2 35012. The two Ivatts had then headed the train from Penrith to Workington. The tour participants boarded a DMU set for a trip to Sellafield and then on to Carlisle while the Ivatts headed to Carlisle with the empty stock. The locomotives had been cleaned with freshly painted smokeboxes and buffer beams for the occasion.

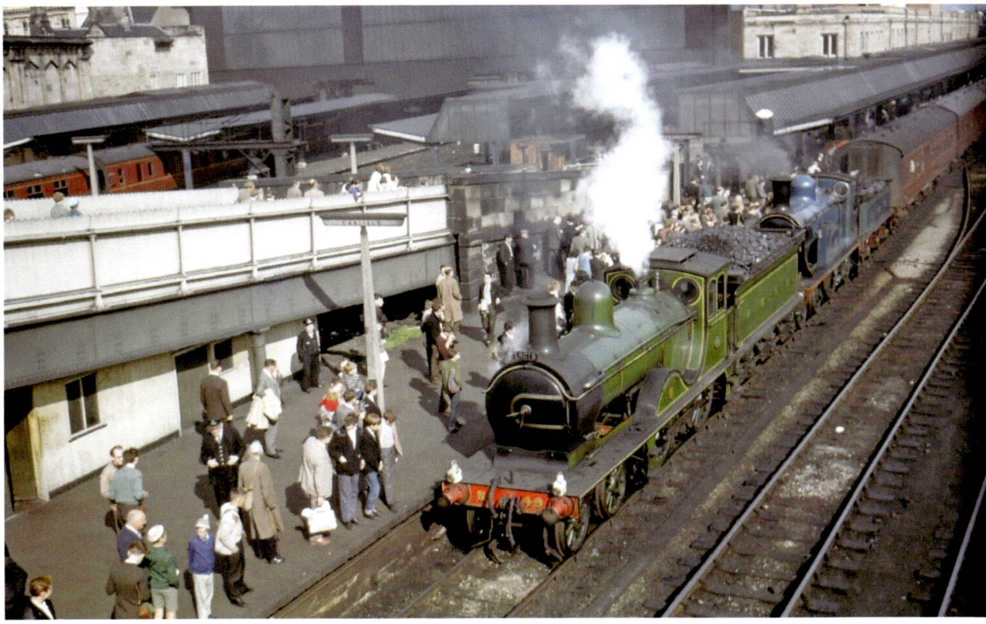

13 June 1964 GNSR 4-4-0 49 and CR 4-2-2 123 at Carlisle Citadel station.
4-4-0 49 was built by the North British Locomotive Company of Glasgow for the Great North of Scotland Railway in October 1920. It had been withdrawn from active service in June 1958 but retained for preservation and returned to steam along with Caledonian Railway 4-2-2 123 in 1959. Both locomotives are seen at Carlisle Citadel station at the head of the 'Solway Ranger'. They were about to depart north from Carlisle for a trip up the Silloth branch and back.

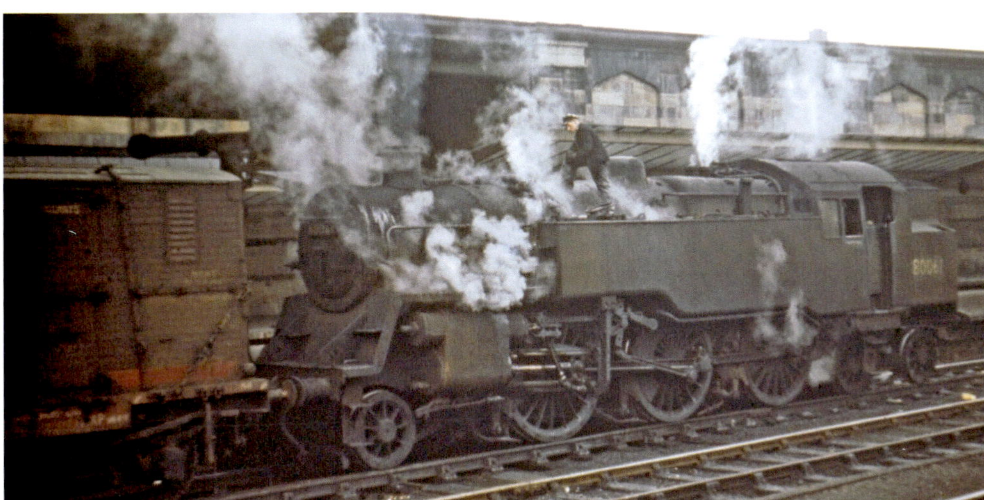

September 1964 BR Standard 4MT 2-6-4T 80061 takes water at Carlisle Citadel.
BR standard 4MT 2-6-4T 80061 had been built at Brighton Works in April 1953. It was first allocated to Bedford MPD and several depots in the North West before its reallocation to Fort William in April 1960. It spent a year at Dumfries MPD from August 1964 and became a regular visitor to Carlisle but moved on to Polmadie depot in 1965. It was withdrawn from there in December 1966.

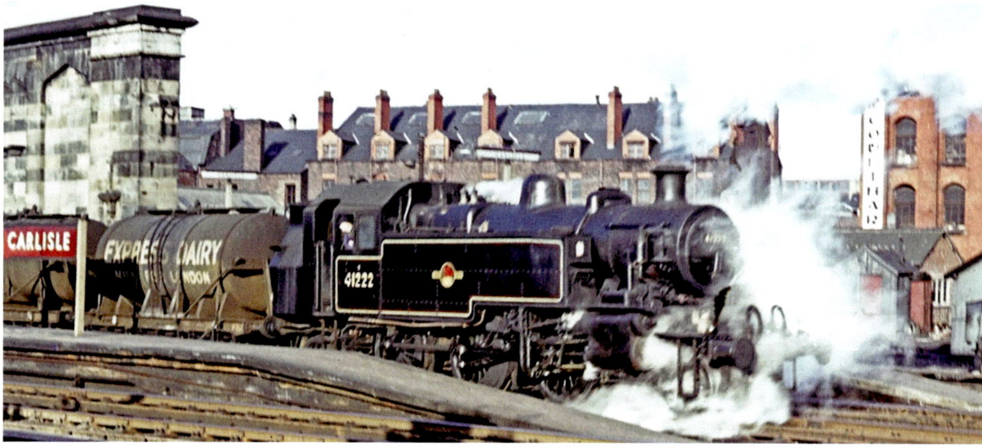

13 February 1965 LMS Ivatt 2MT 2-6-2T 41222 shunting milk tanks at Carlisle Citadel.
LMS Ivatt 2MT 2-6-2T 41222 had been built at Crewe in October 1948 for British Railways. After a year based at Bangor it moved on to Bletchley MPD and remained loyal to that depot until its reallocation to Carlisle Upperby in January 1965. It undertook station pilot and pick-up goods duties until December 1966, after which it was withdrawn from service and scrapped by Shipbreaking Industries of Faslane.

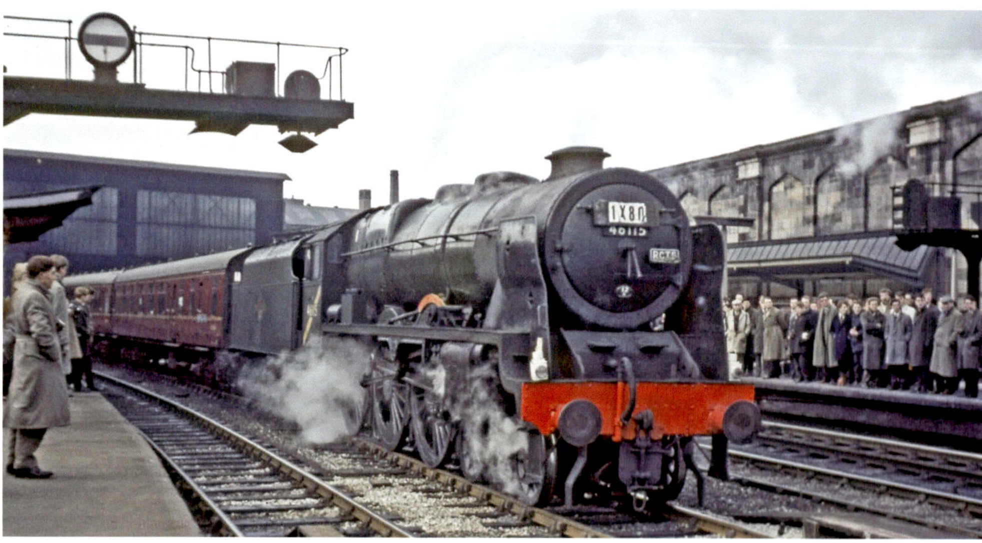

13 February 1965 LMS Royal Scot 4-6-0 46115 shunts stock at Carlisle Citadel.
LMS Royal Scot 4-6-0 46115 *Scots Guardsman* had been built by the North British Locomotive Company of Glasgow for the LMS in October 1927. It was rebuilt in 1947 with a tapered boiler and double chimney. After the rebuild it had been a long-term resident of Manchester's Longsight depot as well as spending time allocated to Crewe North and Springs Branch, Wigan. It was reallocated to Carlisle Upperby in July 1964 and moved across to Kingmoor in November of that year. It was finally withdrawn in January 1966 and purchased for preservation by the late Bob Bill and moved to Haworth on the Keighley Worth Valley Railway. It was seen at Carlisle Citadel at the head of the Rebuilt Scot Commemorative Rail Tour organised by the RCTS which ran between Crewe and Carlisle running via Blackburn on the northward leg to include part of the Settle–Carlisle line.

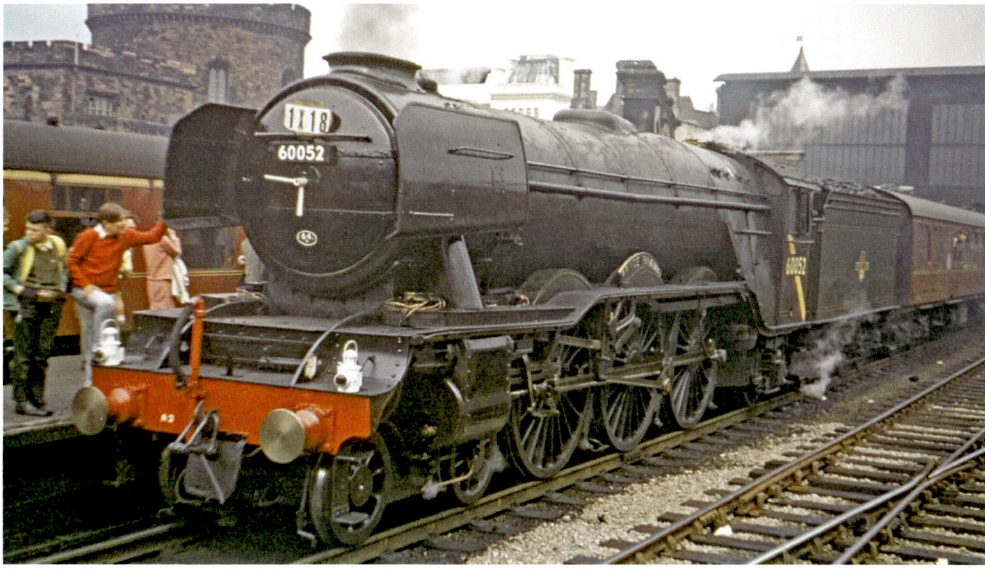

5 June 1965 LNER A3 4-6-2 60052 at Carlisle Citadel.
LNER A3 4-6-2 60052 *Prince Palatine* had been built at Doncaster Plant in November 1924 and had spent much of its career based at depots along the East Coast Main Line. At the formation of British Railways, it had become a resident of ex-Great Central sheds. By August 1955 it had returned to more familiar territory of the North Eastern Region. During this period, it had two modifications: in 1958 it received a double chimney and in 1962 it was fitted with trough-type smoke deflectors. 60052 was the last A3 in BR service and was finally withdrawn in January 1966. It was seen at Carlisle Citadel with the return leg of the Scottish Locomotive Preservation Fund charter rail tour from Edinburgh to Carlisle and back.

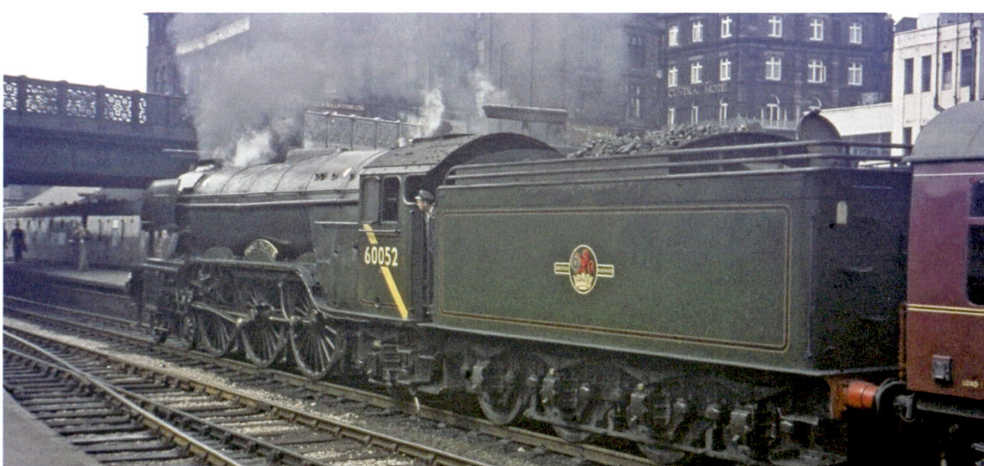

5 June 1965 LNER A3 4-6-2 60052 waits at Carlisle Citadel.
LNER A3 4-6-2 60052 *Prince Palatine* at the head of the return leg of the Scottish Locomotive Preservation Fund charter rail tour from Carlisle to Edinburgh. Although it looks as though the train is about to depart, 60052 was taken off the train due to a hot trailing wheel axle-box. Fairly quickly A4 60027 *Merlin*, in appalling external condition, was substituted and returned the train to Edinburgh.

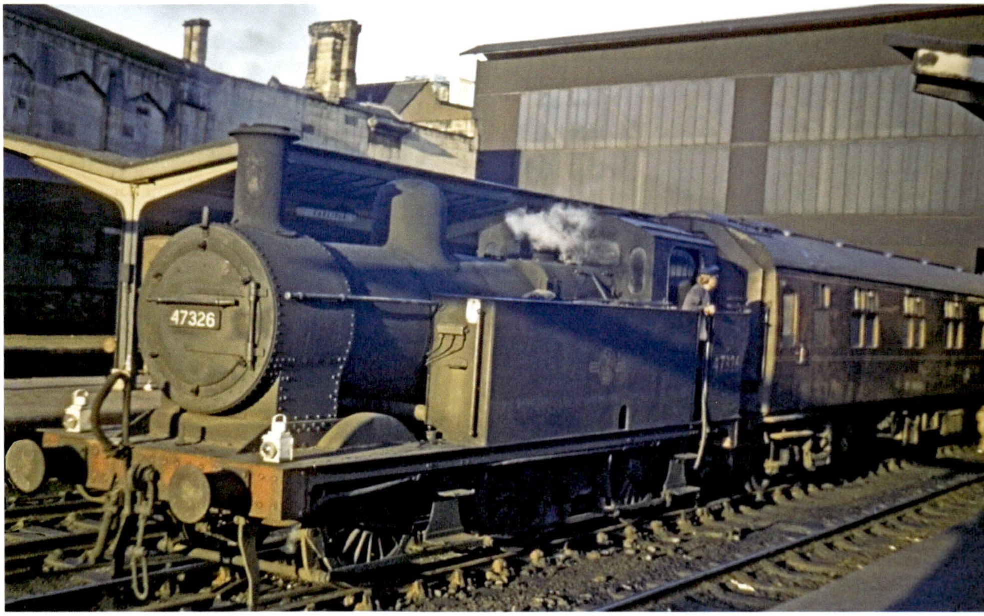

June 1965 LMS Jinty 3F 0-6-0T 47326 shunting at Carlisle Citadel.
LMS Jinty 3F 0-6-0T 47326 was built by the North British Locomotive Company of Glasgow for the LMS in July 1926 and at the formation of British Railways was allocated to Carlisle Upperby MPD. It was transferred across to Kingmoor depot in April 1965 and was withdrawn in December 1966. It was eventually cut up by McWilliam of Shettleston.

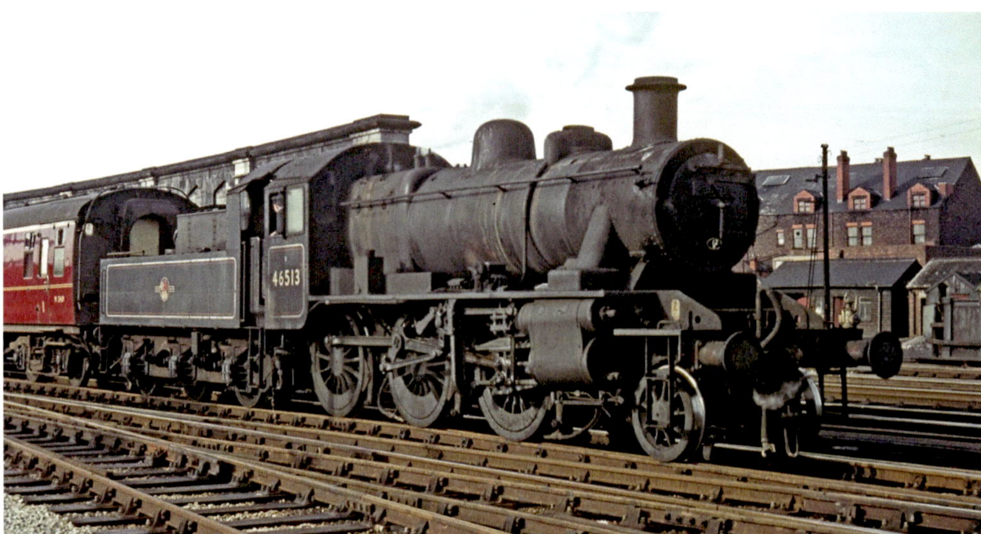

June 1965 LMS Ivatt 2MT 2-6-0 46513 shunting coaching stock at Carlisle Citadel.
LMS Ivatt 2MT 2-6-0 46513 was built for British Railways at Swindon Works in December 1952. It was allocated to Oswestry MPD from new but moved to Newton Heath in February 1965. After only five months it found its final home at Carlisle Upperby, arriving there in July 1965. It was withdrawn after serving the depot for a year and was cut up by Motherwell Machinery and Scrap of Wishaw.

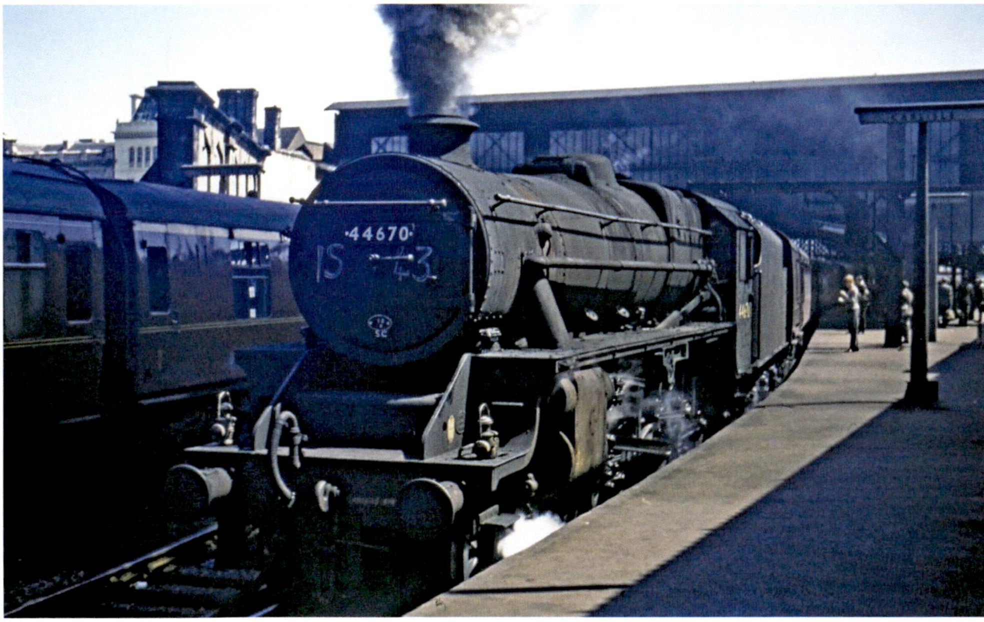

August 1965 LMS Black Five 4-6-0 44670 at Carlisle Citadel.
LMS Black Five 4-6-0 44670 had been built for British Railways by Horwich Works and was part of the last batch of twenty-six of the class built in 1950. It had been built with Skeflo bearings on its driving wheel axles. There were two more Black Fives built in 1951 but these featured Caprotti valve gear and high running plates. 44670 spent much of its career based at Carlisle Kingmoor and was withdrawn from the depot in January 1966. It was cut up by T. W. Ward of Beighton.

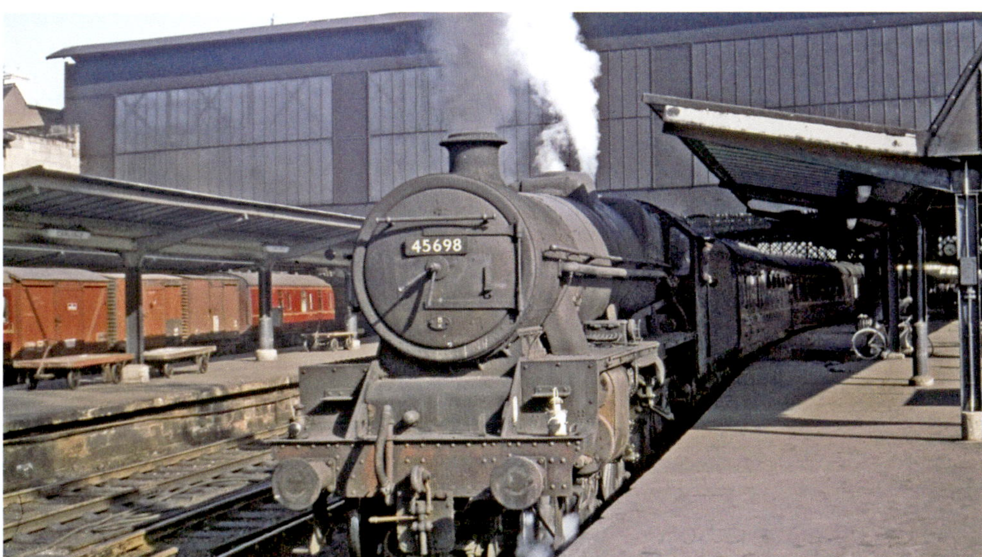

13 August 1965 LMS Jubilee 4-6-0 45698 awaits departure at Carlisle Citadel.
LMS Jubilee 4-6-0 45698 *Mars* had been built at Crewe in April 1936 and spent its early years of service in the North West. In May 1948 it was transferred to Southport MPD and it was to remain a constant at this depot until its withdrawal in February 1966.

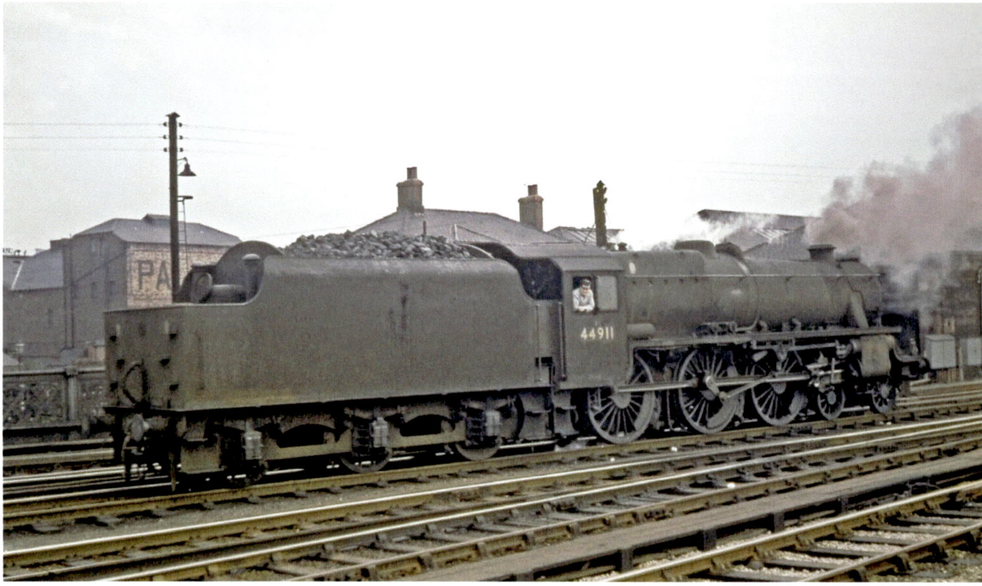

15 August 1965 LMS Black Five 4-6-0 44911 at Carlisle Citadel.
LMS Black Five 4-6-0 44911 was built at Crewe in November 1945 and had been allocated to numerous depots throughout its career, most notably spending fifteen months at King's Cross MPD from February 1956. In June 1964 it was transferred to Carlisle, the first five months at Upperby depot before being moved to Kingmoor. It was withdrawn in October 1967 and finally cut up by McWilliam of Shettleston in May 1968.

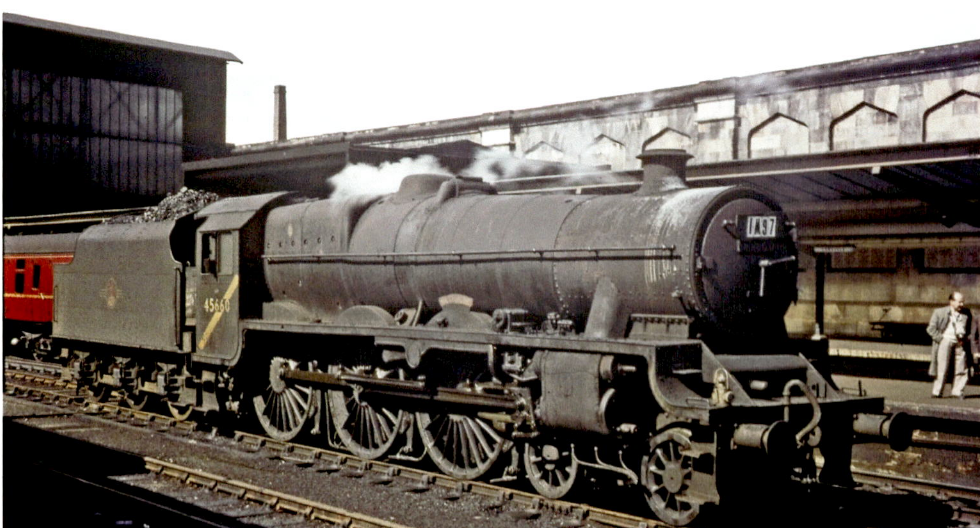

September 1965 LMS Jubilee 4-6-0 45660 at Carlisle Citadel.
LMS Jubilee 4-6-0 45660 *Rooke* was built at Derby Works in December 1934. It began its British Railways career at Holbeck MPD but was quickly transferred away to Bristol Barrow Road in May 1948 and then on to Shrewsbury in October 1961. In September 1964 it found its way back to Holbeck and was withdrawn from there in June 1966.

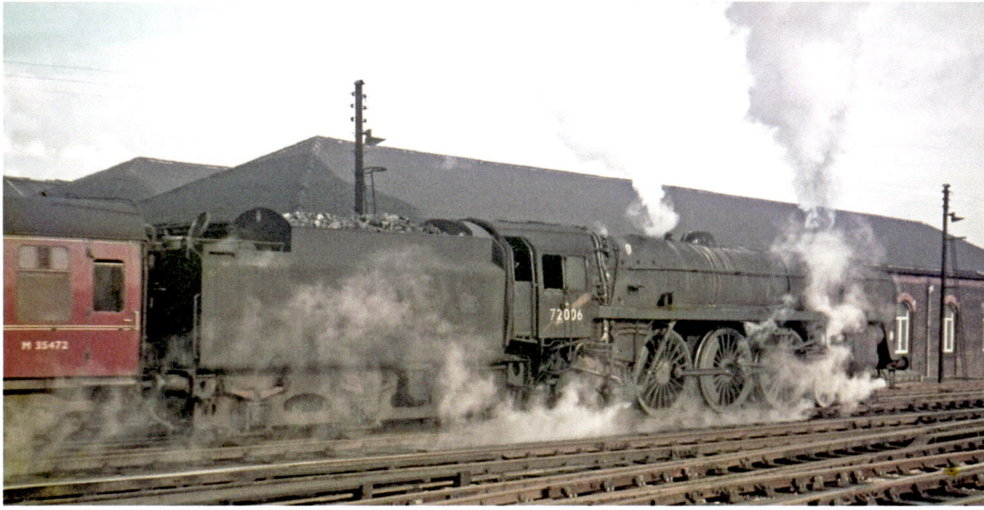

September 1965 BR Standard Clan 6MT 4-6-2 72006 departs southbound from Carlisle Citadel.
BR Standard Clan 6MT 4-6-2 72006 *Clan Mackenzie* had been built at Crewe Works in February 1952 and was allocated to Carlisle Kingmoor MPD. Other than a six-month sojourn at Haymarket depot in 1958, it remained faithful to Kingmoor until its withdrawal in May 1966. It was cut up in October 1966 by J. McWilliam of Shettleston.

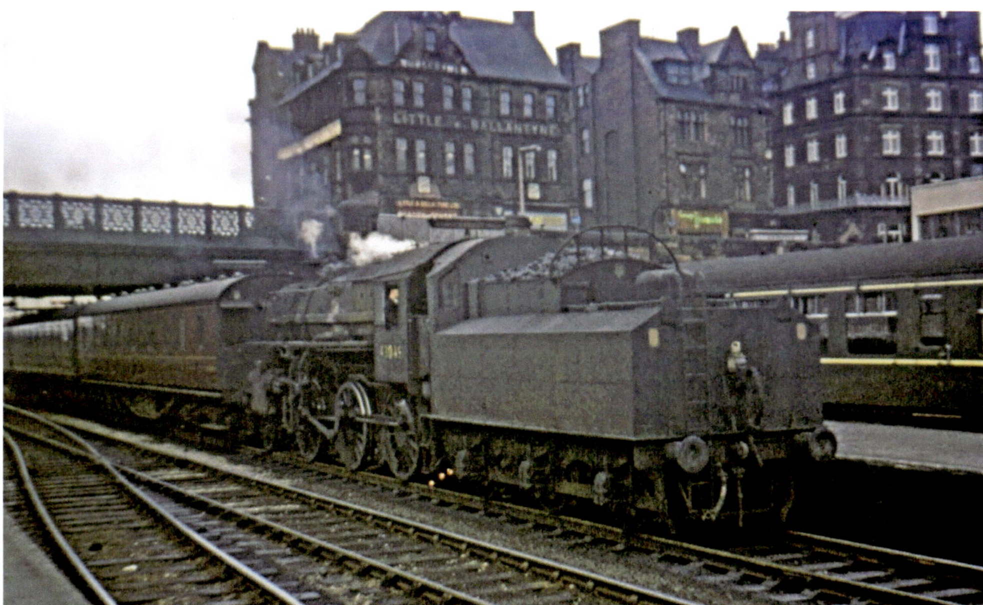

23 April 1966 LMS Ivatt 4MT 2-6-0 43049 arrives at Carlisle Citadel.
LMS Ivatt 4MT 2-6-0 43049 had been built at Horwich Works for British Railways in November 1949 being allocated new to Derby MPD. In August 1956 it was transferred to Saltley depot and another move in September 1962 took it to Heaton Mersey. It arrived on the roster at Carlisle Kingmoor in November 1964 and remained active there until withdrawal in August 1967. It was cut up by the Motherwell Machinery and Scrap Company in Wishaw. It was seen arriving at Carlisle Citadel from Langholm.

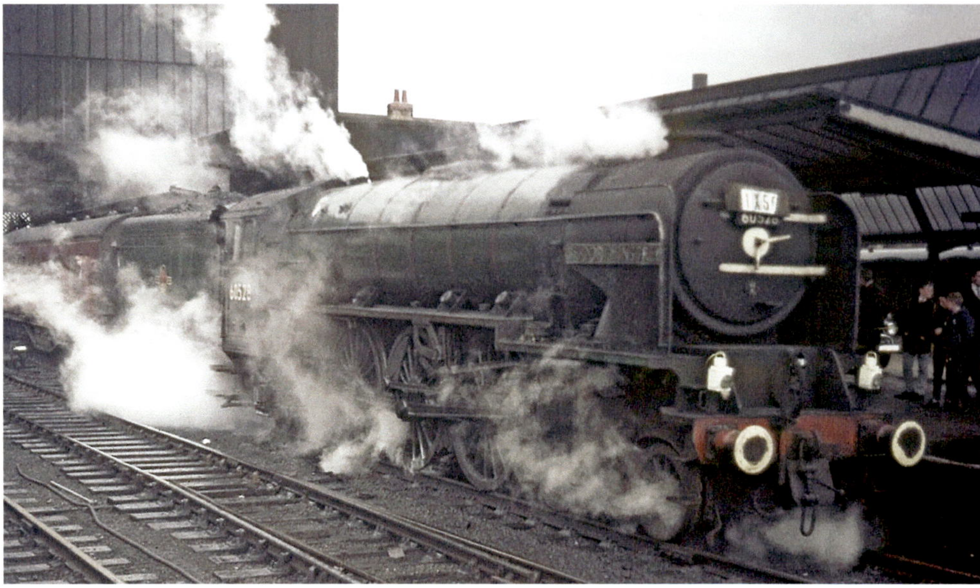

23 April 1966 LNER A2 4-6-2 60528 at Carlisle Citadel.
LNER A2 4-6-2 60528 *Tudor Minstrel* was built at Doncaster Plant in February 1948 and was sent to Gateshead MPD. The following year it was reallocated to Dundee Tay Bridge depot and remained a Scottish-based locomotive until it was withdrawn in June 1966. During its last few months of service it was allocated to Aberdeen Ferryhill MPD and was a regular performer on enthusiast rail tours. It was eventually cut up by J. McWilliam of Shettleston.

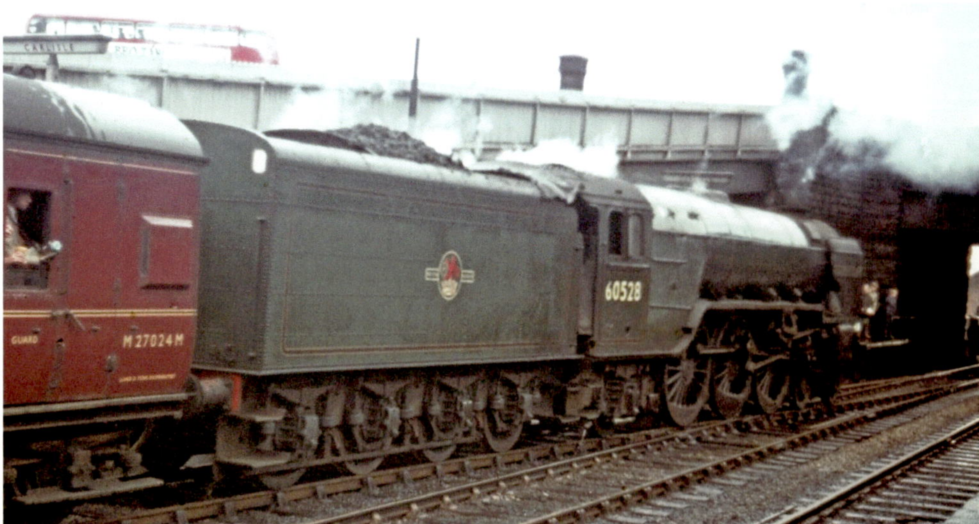

23 April 1966 LNER A2 4-6-2 60528 departs from Carlisle Citadel.
LNER A2 4-6-2 60528 *Tudor Minstrel* was seen departing north from Carlisle Citadel to take the Waverley Route to Edinburgh at the head of the 'Waverley Special' organised by the Altrinchamian Railway Excursion Society. The locomotive worked the train from Manchester Exchange via Wigan and Carlisle through to Edinburgh. The return journey was via Berwick and York using two V2s and 45565.

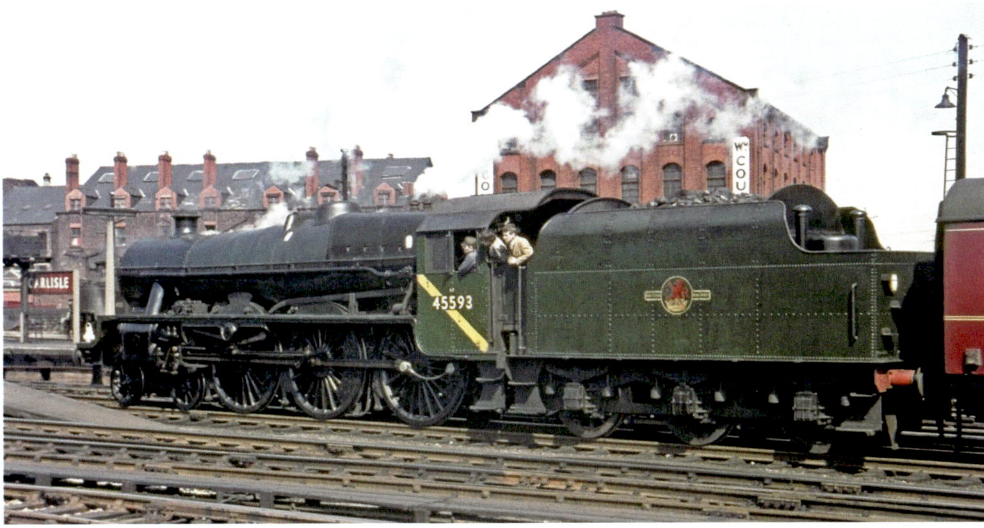

30 April 1966 LMS Stanier Jubilee 4-6-0 45593 enters Carlisle Citadel.

LMS Jubilee 4-6-0 45593 *Kolhapur* was in its final years of service based at Holbeck MPD. It was regularly used throughout that summer between Leeds and Carlisle on relief trains to Glasgow. It was also kept busy working specials and rail tours at a time when the days of steam on the main line were numbered. At the end of the summer timetable in 1967 its working days were over and it was purchased for preservation by the then Standard Gauge Steam Trust. It has remained fairly active on the main line and on heritage railways over the intervening years. It now awaits another overhaul at the Tyseley Locomotive Works to put it back in into steam.

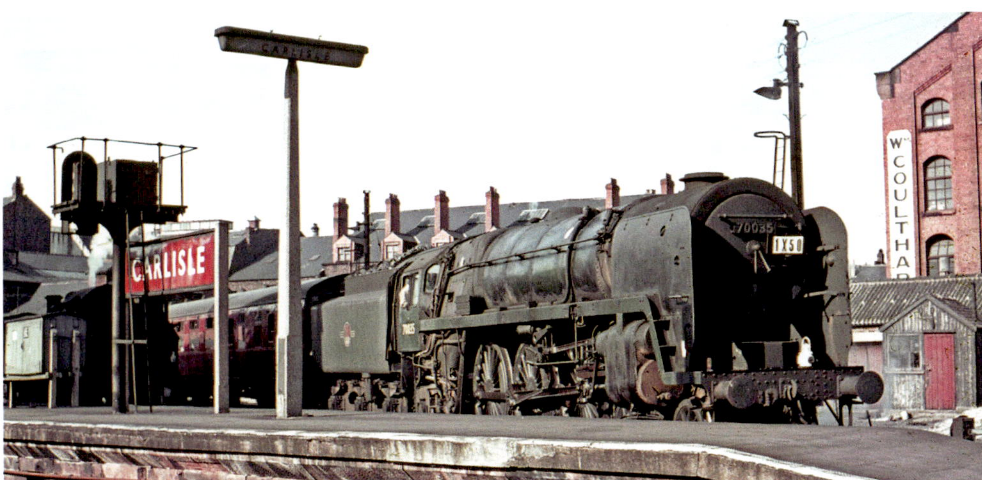

30 April 1966 BR Standard Britannia 7MT 4-6-2 70035 at Carlisle Citadel.

BR Standard Britannia 7MT 4-6-2 70035 *Rudyard Kipling* had been built at Crewe Works in December 1952. It was based at Norwich Thorpe MPD and began work on the Great Eastern main line from Norwich to Liverpool Street station, London. When this route was dieselised it was transferred to Immingham depot along with six other Britannias in October 1961. By December 1963 it had been transferred to Carlisle Kingmoor and remained active there until withdrawal in December 1967. It was cut up by T. W. Ward's of Inverkeithing in the spring of 1968. (Photograph L. Flint)

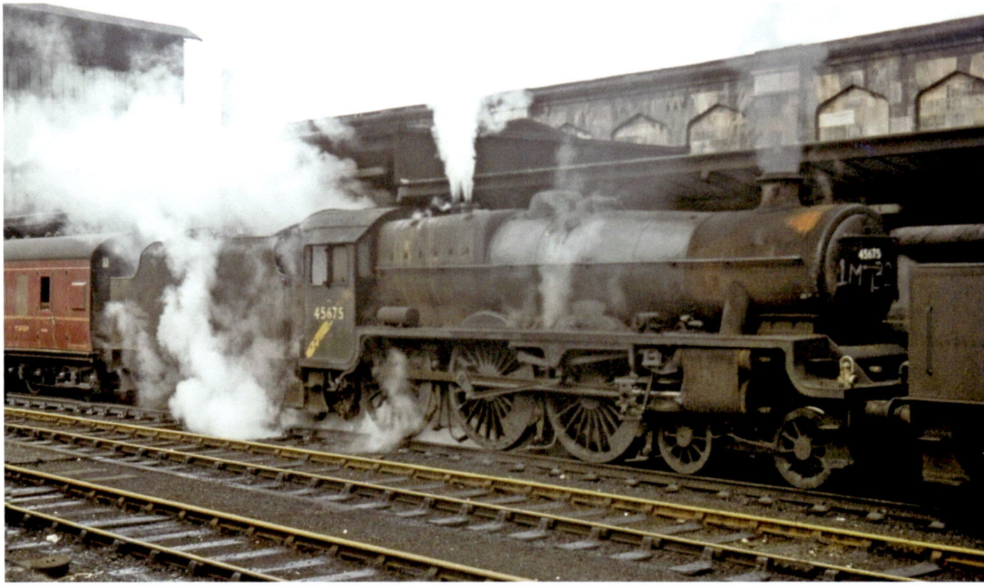

July 1966 LMS Jubilee 4-6-0 45675 at Carlisle Citadel.
LMS Jubilee 4-6-0 45675 *Hardy*, ably assisted by LMS Black Five 4-6-0 45490, gets its train on the move again after a pause at Carlisle Citadel. 45675 had been built at Crewe Works in December 1935 and had been based at Carlisle Upperby when new. It had spent all its British Railways career allocated to Holbeck MPD, Leeds, and was withdrawn from there in June 1967. Its last journey was to Cashmore's of Great Bridge for cutting up in December 1967.

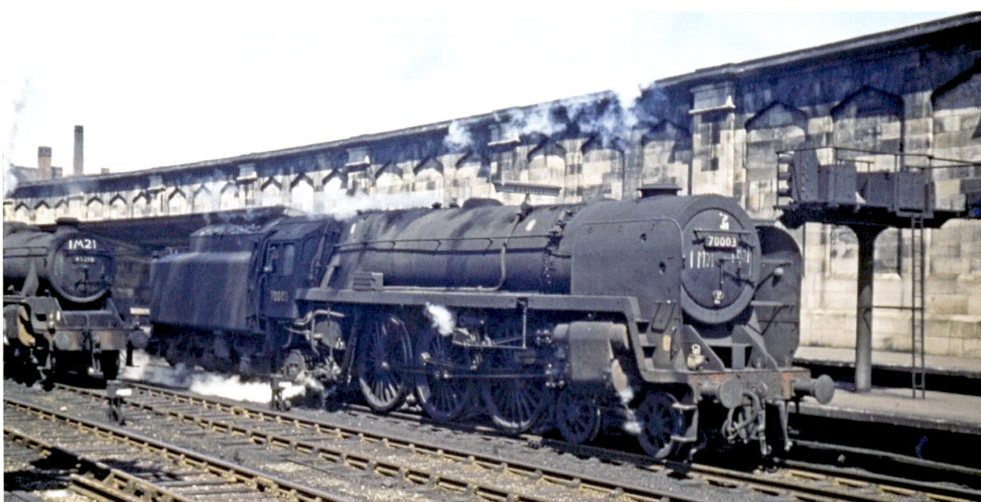

July 1966 BR Standard Britannia 7MT 4-6-2 70003 at Carlisle Citadel.
BR Standard Britannia 7MT 4-6-2 70003 *John Bunyan* had been built at Crewe Works and entered service in March 1951 allocated to Stratford MPD. In July 1961 it became a Norwich Thorpe locomotive but moved again in December 1963 to Carlisle Kingmoor from where it was withdrawn in March 1967. It was cut up by Campbell's of Airdrie in November 1967. It had arrived at Carlisle from the north with a summer extra to Blackpool and was in the process of being replaced by the Black Five for the final leg of the journey.

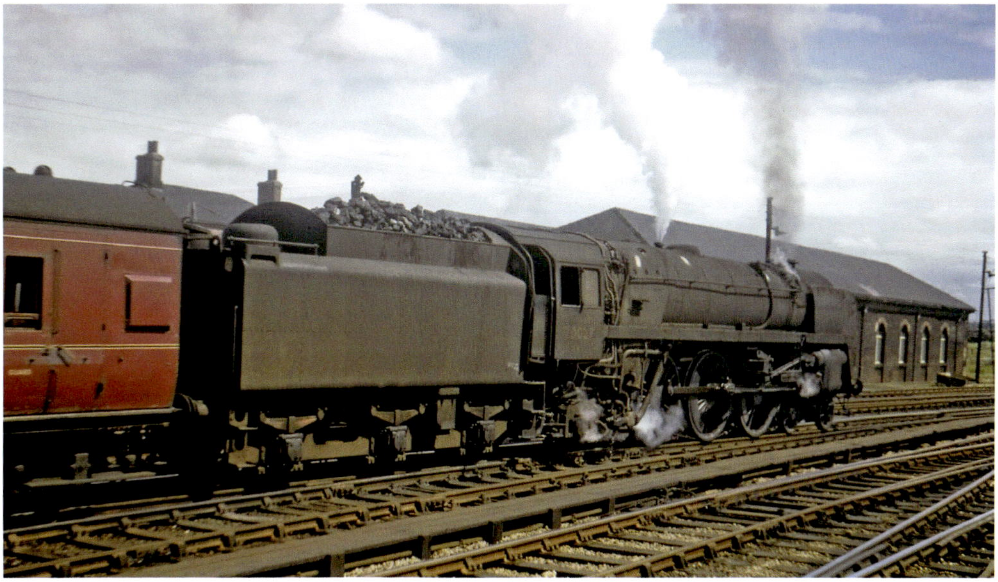

July 1966 BR Standard Britannia 7MT 4-6-2 70027 at Carlisle.
BR Standard Britannia 7MT 4-6-2 70027 *Rising Star* had emerged from Crewe Works in October 1952. It was allocated to Cardiff Canton to work the South Wales express services to London Paddington. In September 1961 it moved on to Aston MPD and September 1966 saw it at its final home of Carlisle Kingmoor from where it was withdrawn in in July 1967.

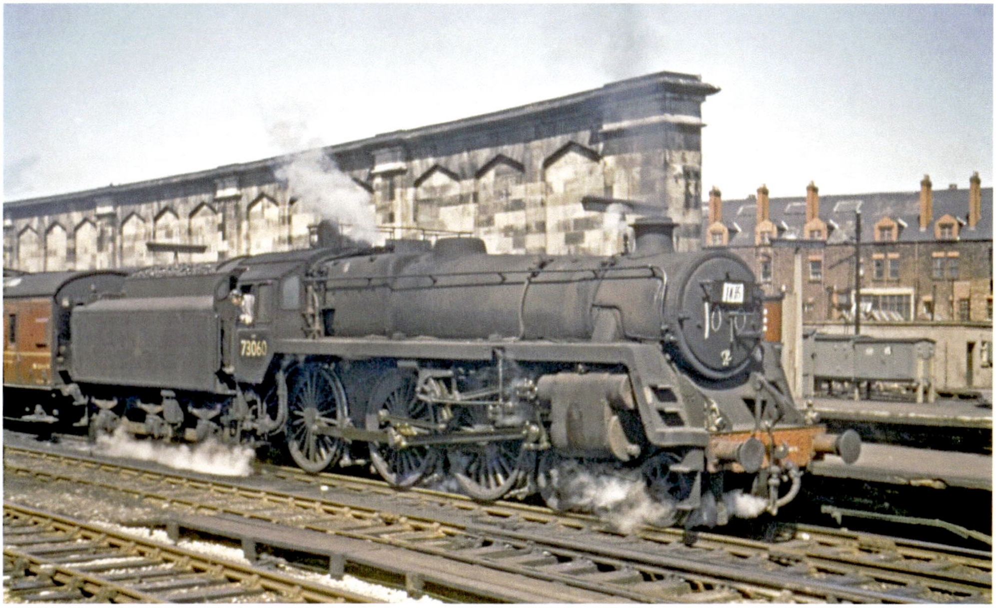

July 1966 BR Standard 5MT 4-6-0 73060 at Carlisle Citadel.
BR Standard 5MT 4-6-0 73060 was built at Derby Works in August 1954 and went to Glasgow Polmadie depot from new. Other than two years spent at Motherwell it remained faithful to its first allocation. It was withdrawn in May 1967 and cut up by Campbell's of Airdrie in September 1967.

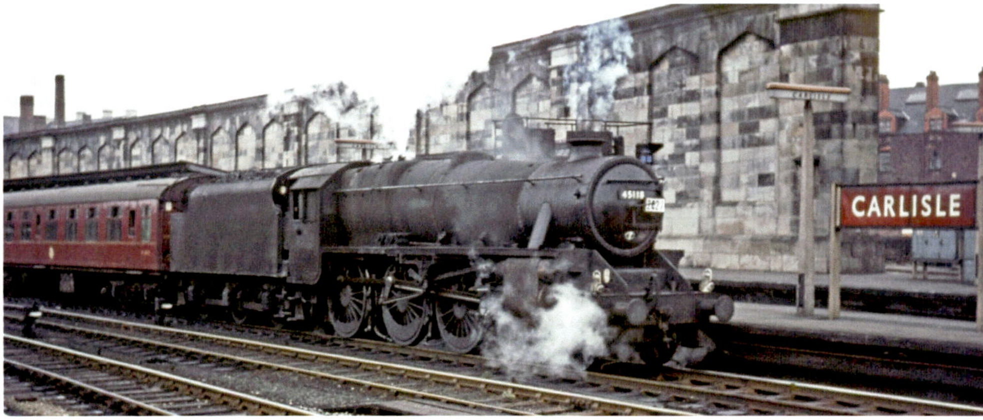

23 July 1966 LMS Black Five 4-6-0 45118 departs Carlisle Citadel with a Glasgow–Blackpool working.

LMS Black Five 4-6-0 45118 had been built by the Vulcan Foundry for the LMS in June 1935. It had mainly been based at Carlisle Kingmoor during its tenure with British Railways. It did spend fifteen months allocated to Perth and a few months at Aintree and Newton Heath, but returned to Carlisle for the rest of its active service. It was withdrawn in October 1966 and was cut up by Draper's of Hull in March 1967.

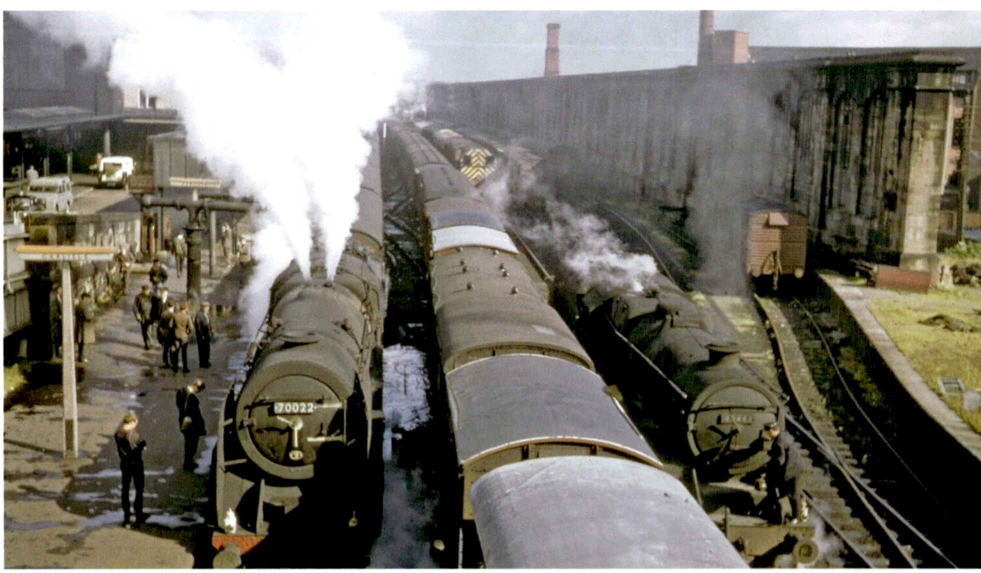

30 July 1966 BR Standard Britannia 7MT 4-6-2 70022 and LMS Black Five 4-6-0 45463 at Carlisle Citadel.

BR Standard Britannia 7MT 4-6-2 70022 *Tornado* was built at Crewe Works in August 1951. It spent the first five years of service working from Plymouth Laira and Newton Abbot before being transferred to Cardiff Canton in 1956. In September 1961 it moved to the London Midland Region, first to Carlisle Kingmoor and then spent time at Longsight, Aston and Rugby depots before returning to Carlisle Upperby in November 1964. With the closure of Upperby depot to steam in December 1966 it moved across to Kingmoor for its final year in traffic. It was seen at the north end of Carlisle Citadel station from the Victoria Viaduct and was about to get a Manchester to Glasgow train on the move.

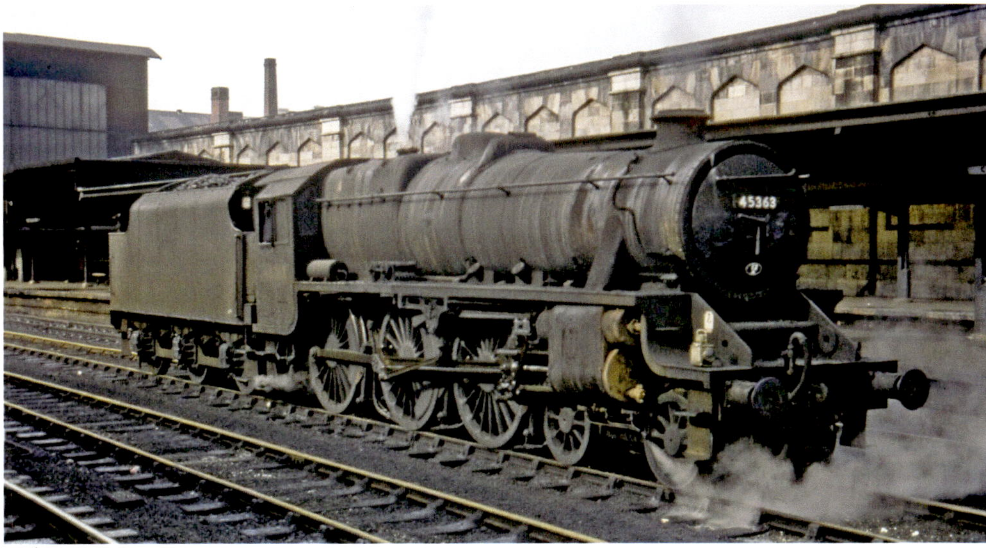

August 1966 LMS Black Five 4-6-0 45363 waits in Carlisle Citadel.
LMS Black Five 4-6-0 45363 was a product of Armstrong Whitworth of Newcastle for the LMS in June 1937. It had been allocated to Carlisle Kingmoor MPD before the formation of British Railways in January 1948. It stayed faithful to that depot until withdrawal in October 1967. It became another Kingmoor resident cut up by J. McWilliam of Shettleston in April 1968.

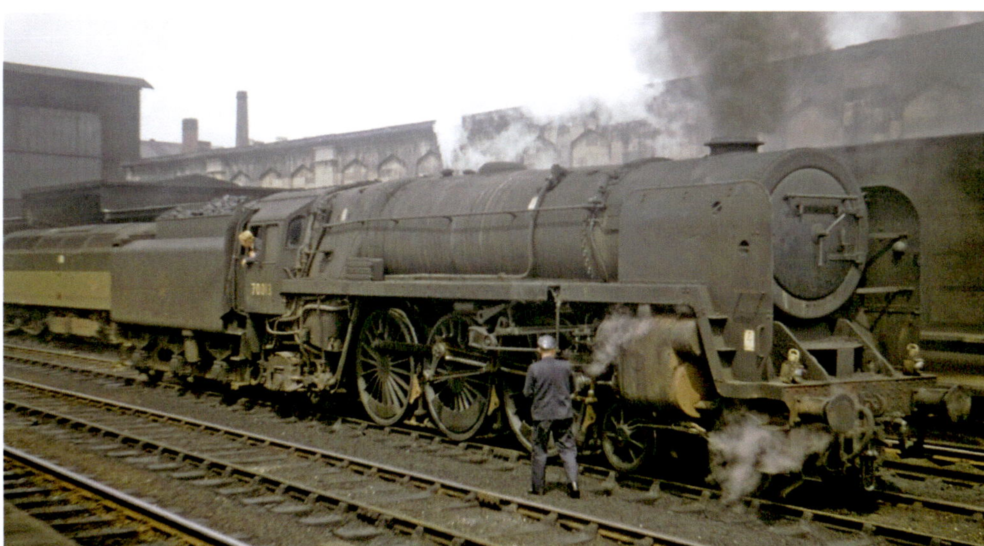

August 1966 BR Standard Britannia 7MT 4-6-2 70013 in Carlisle Citadel.
BR Standard Britannia 7MT 4-6-2 70013 *Oliver Cromwell* emerged from Crewe Works in May 1951. It was at first allocated to Norwich Thorpe MPD, later to Ipswich and then to March. In December 1963 it left the Great Eastern Region for Carlisle and served at both Upperby and Kingmoor. With the closure of Kingmoor it was transferred to Carnforth in January 1968. It was seen in work-stained condition without nameplates or smokebox numberplate; it was to receive a general overhaul at Crewe Works in early 1967 and was to take a leading role in the '15 Guinea Special' on the last day of steam on British Railways after which it went into preservation.

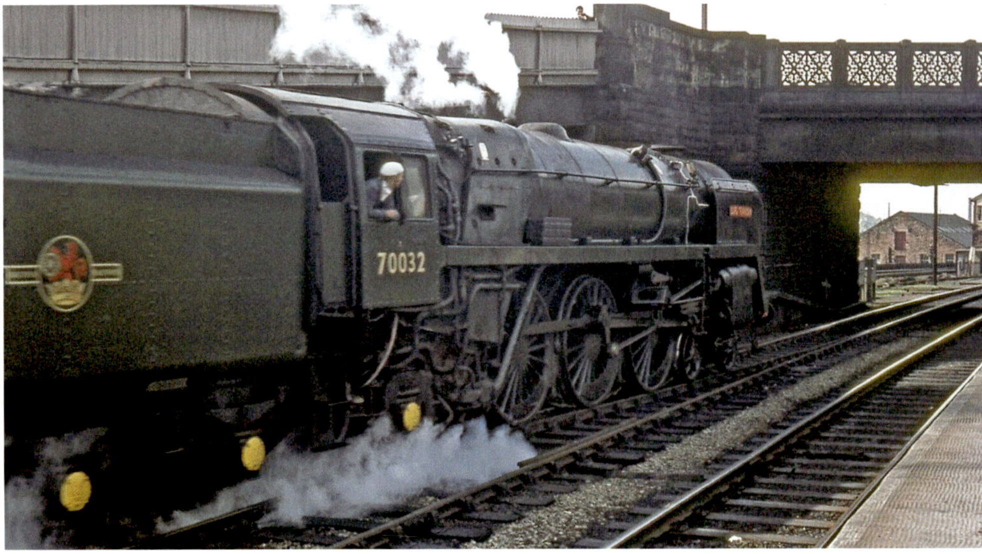

3 September 1966 BR Britannia 4-6-2 70032 arrives at Carlisle with the Granite City Rail Tour.
BR Standard Britannia 7MT 4-6-2 70032 *Tennyson* stands at Carlisle Citadel having arrived with the South and West Railway Society's 'Granite City Special'. It was an ambitious special which began at London Euston and travelled to Aberdeen via the West Coast Main Line, returning to London King's Cross via the East Coast Main Line. One electric, six steam and one diesel were used to haul the train over various sections. 70032 had hauled the train from Preston to Carlisle.

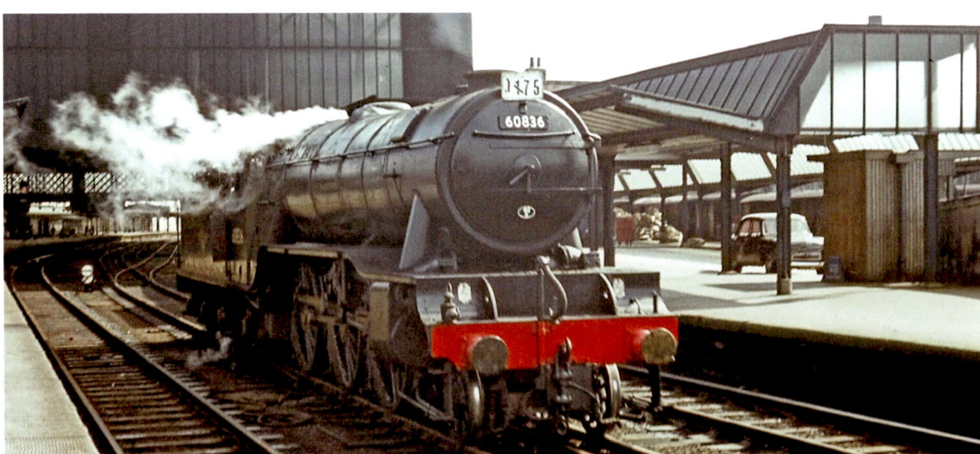

3 September 1966 LNER 4-6-2 V2 60836 at Carlisle Citadel.
LNER 4-6-2 V2 60836 had been built at Darlington Works in September 1938. It had been a Scottish-based locomotive throughout its entire career, which began and ended at Dundee Tay Bridge depot. It had been resident at a number of other depots over the years, particularly Haymarket and St Margaret's. It was the last active V2 on British Railways and after its withdrawal it served as a steam-heating unit at Craigentinny Carriage sidings until April 1967. After four months in store at Thornton Junction it was finally cut up by Motherwell Machinery and Scrap of Wishaw. It was seen at Carlisle Citadel as it awaited the arrival of the South and West Railway Society's Granite City Rail Tour which it would take from Carlisle to Edinburgh over the Waverley route.

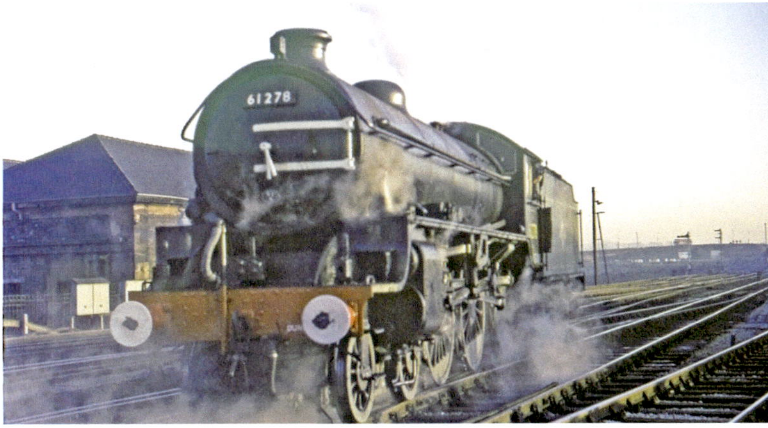

3 December 1966 LNER B1 4-6-0 61278 at Carlisle Citadel.
LNER B1 4-6-0 61278 manoeuvres at Carlisle Citadel while working the Last B1 Excursion. The special had been organised by the Scottish Region. It ran between Edinburgh and Carlisle Kingmoor MPD and included a depot visit. The outward route from Edinburgh was via Airdrie, Kilmarnock and Dumfries with the return journey taking the Waverley route. 61278 had been built by the North British Locomotive Company of Glasgow for the LNER, but was actually delivered in January 1948, to the newly formed British Railways. It had been allocated to Dundee Tay Bridge MPD throughout its career. It was withdrawn from there in April 1967 and cut up in January 1968 by Arnott Young of Carmyle.

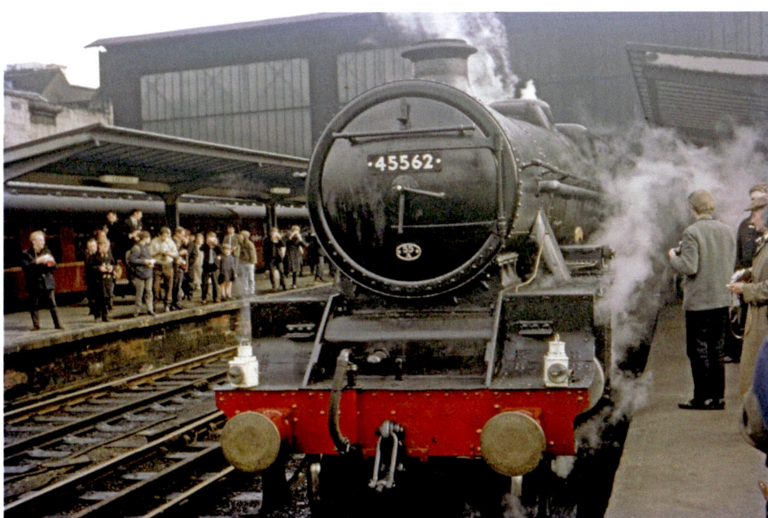

25 February 1967 LMS Jubilee 4-6-0 45562 at Carlisle Citadel.
LMS Stanier Jubilee 4-6-0 45562 *Alberta* had been a Leeds-based locomotive for much of its career and had become a 'celebrity' locomotive by the start of 1967. It was one of a trio of Holbeck-based survivors of the class to receive a lot of attention from shed staff and enthusiasts to keep them looking immaculate. The engines were in high demand throughout the summer to haul enthusiast specials and rail tours, as well as undertaking important express passenger duties over the Settle and Carlisle line. It had arrived at Carlisle after hauling the Jubilee Locomotive Preservation Society's special, 'The Border Countryman', from Leeds via Preston and the West Coast Main Line. It was to return the train back to Leeds via Newcastle and the East Coast Main Line to York.

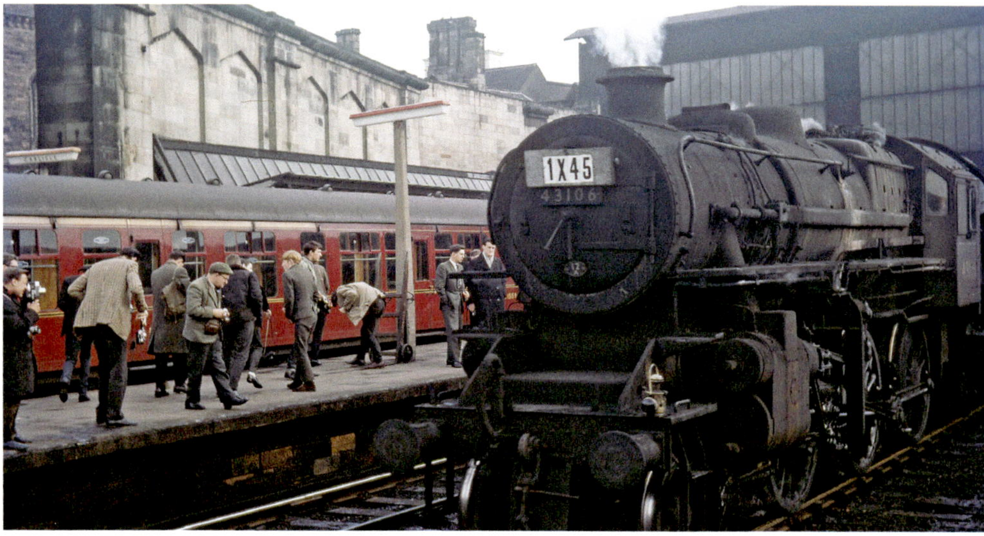

25 February 1967 LMS Ivatt 4MT 2-6-0 43106 at Carlisle.
LMS Ivatt 4MT 2-6-0 43106 became the centre of the photographers' attention as it prepared to attach to the Jubilee Locomotive Preservation Society's special, 'The Border Countryman', which it was to haul from Carlisle to Beattock and back. Unlike Jubilee 4-6-0 45562 *Alberta*, which had hauled the train from Leeds via Preston and up the West Coast Main Line, 43106 was in a very care-worn condition and had not seen any attention from the shed cleaners or local enthusiasts. It is somewhat ironic that the immaculate 45562 was to be cut up before the end of 1967 and 43106 went on to be preserved on the Severn Valley Railway. It is still regularly at work over fifty-five years later.

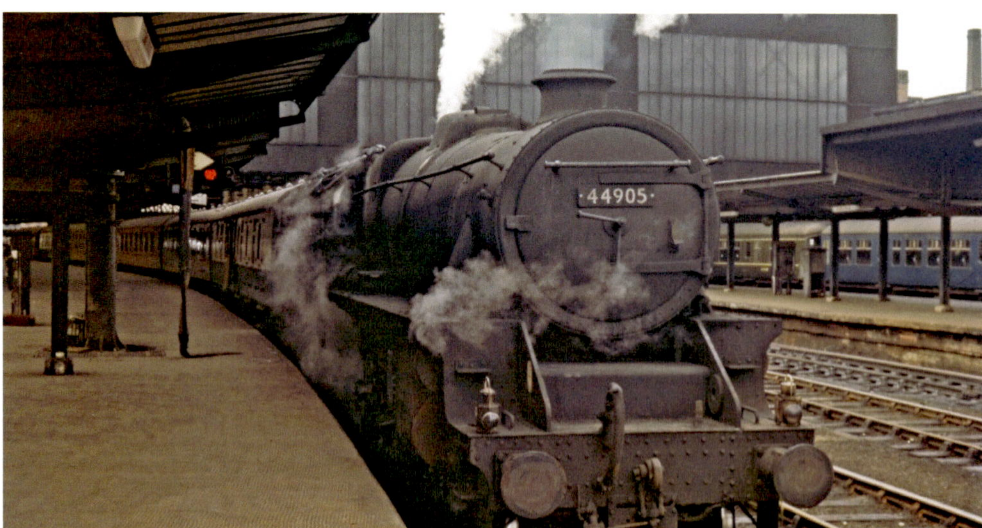

April 1967 LMS Black Five 4-6-0 44905 awaits departure from Carlisle Citadel.
LMS Black Five 4-6-0 44905 had been built at Crewe in October 1945. It had spent its entire career based at depots in the North West, most notably Carnforth, where it was resident from October 1950. It was eventually withdrawn from Carnforth towards the end of steam on British Railways in late 1967. It was cut up by Draper's of Hull in February 1968.

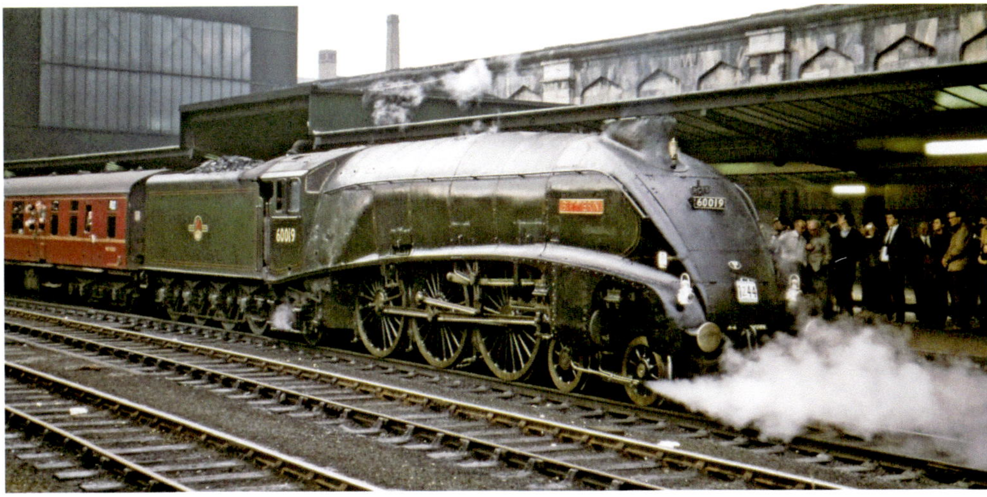

16 July 1967 LNER A4 4-6-2 60019 at Carlisle with the RCTS A4 to Glasgow Rail Tour. LNER A4 4-6-2 60019 *Bittern* had been built at Doncaster Plant in December 1937 and spent much of its career based in the North East, at Gateshead and Heaton depots. With the dieselisation of services on the East Coast Main Line its duties disappeared and it was transferred to the Scottish Region in October 1963. It quickly found a home at Aberdeen Ferryhill MPD from where it worked the Aberdeen to Glasgow three-hour express services. 60019 was seen paused at Carlisle Citadel with the RCTS West Riding Branch rail tour A4 to Glasgow, which it headed from Leeds to Glasgow and return.

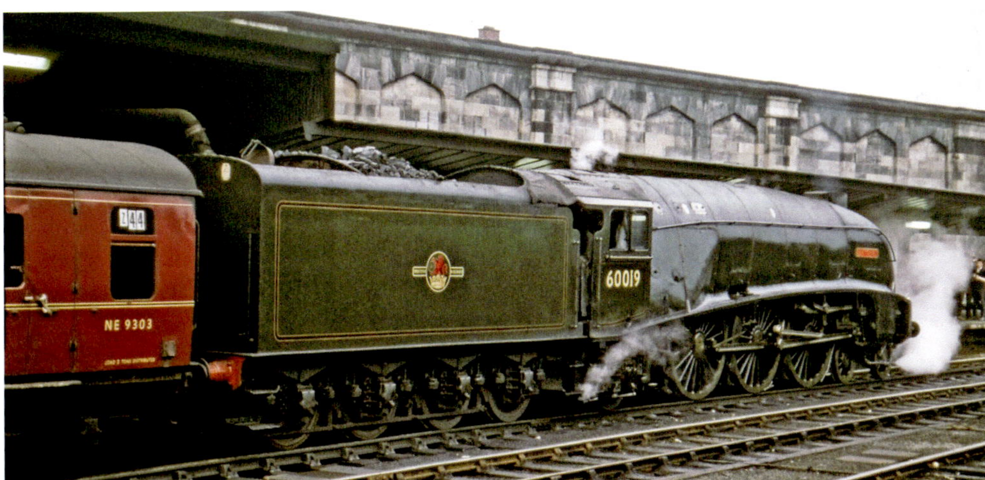

16 July 1967 LNER A4 4-6-2 60019 at Carlisle with the RCTS A4 to Glasgow Rail Tour. LNER A4 4-6-2 60019 *Bittern* had worked from Aberdeen Ferryhill for its last six years in service until withdrawal in September 1966. Upon withdrawal it was purchased privately for preservation by C. L. Drury and based at York MPD. It continued its active life in preservation, working rail tours and specials until major maintenance problems curtailed these workings in the early 1970s. It spent the next thirty-five years in store at various locations including Neville Hill, Walton Colliery, Dinting and Southall. It changed ownership twice in that time but was eventually restored to working order in 2007. It has since performed regularly on the main line and on heritage railways.

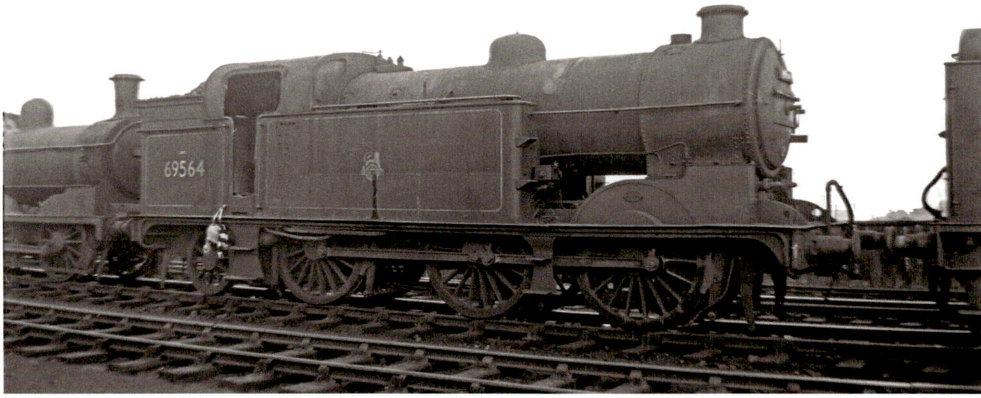

March 1960 LNER N2 0-6-2T 69564 at Carlisle Canal MPD.
LNER N2 0-6-2T 69564 had been built at Doncaster Plant in November 1920 and had spent most of its career working suburban trains around Glasgow. It had been allocated to Parkhead and later Eastfield depots. In August 1957 it was reallocated to Carlisle Canal and was withdrawn from there in June 1961. Afterwards it was stored at Corkerhill MPD before being scrapped at Cowlairs Works in late 1962.

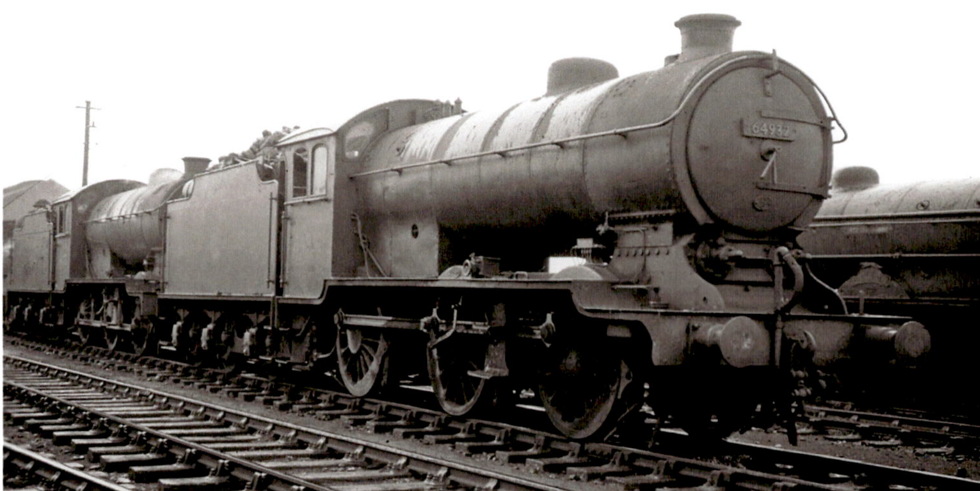

May 1961 LNER J39 0-6-0 64932 at Carlisle Canal MPD.
LNER J39 0-6-0 64932 had been built in May 1937 by Beyer Peacock of Manchester for the LNER. It spent its entire active life allocated to Carlisle Canal and was regularly seen on the Silloth branch and the sub-shed there. It was withdrawn from service in July 1961 and was cut up at Cowlairs Works by the end of that year.

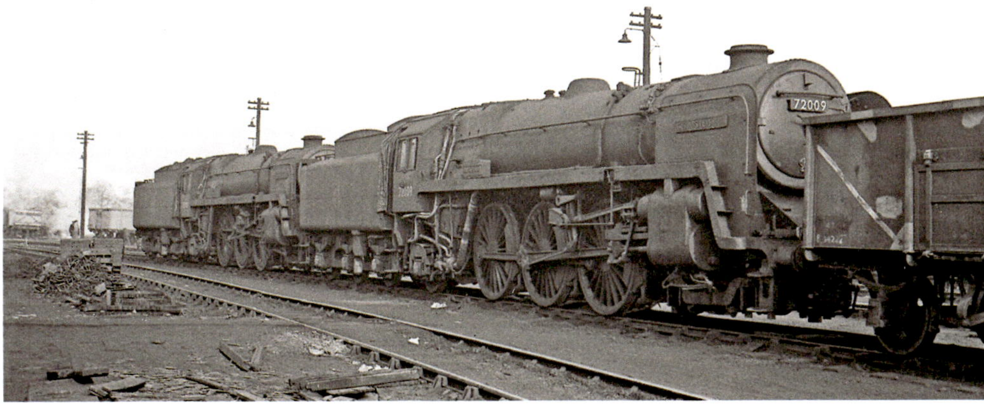

July 1960 BR Standard Clan 6MT 4-6-2s 72009 with 72007 at Carlisle Kingmoor MPD.
BR Standard Clan 6MT 4-6-2s 72009 *Clan Stewart* and 72007 *Clan Mackintosh* followed each other into service at Carlisle Kingmoor in 1952. They both remained at Kingmoor throughout their careers and were seen here with their front bogie wheels removed for some maintenance work. They returned to work shortly after the completion of the repairs.

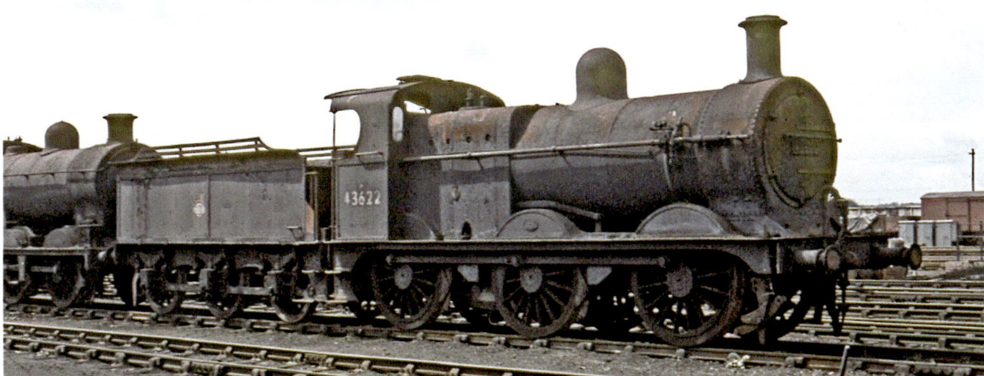

24 June 1962 MR 3F 0-6-0 43622 out of use at Carlisle Kingmoor MPD.
MR 3F 0-6-0 43622 had been built by Neilson, Reid and Company of Glasgow for the Midland Railway in 1899. It had spent a considerable amount of its working life at Hasland depot in Derbyshire, remaining there at the formation of British Railways in January 1948. It was to retain its number and insigna of its previous owners until April 1952 when it eventually received an overhaul. It was reallocated to Carlisle Kingmoor MPD in May 1953 and worked from there until being withdrawn in December 1959. It was then stored out of use and finally cut up at Horwich Works in December 1962.

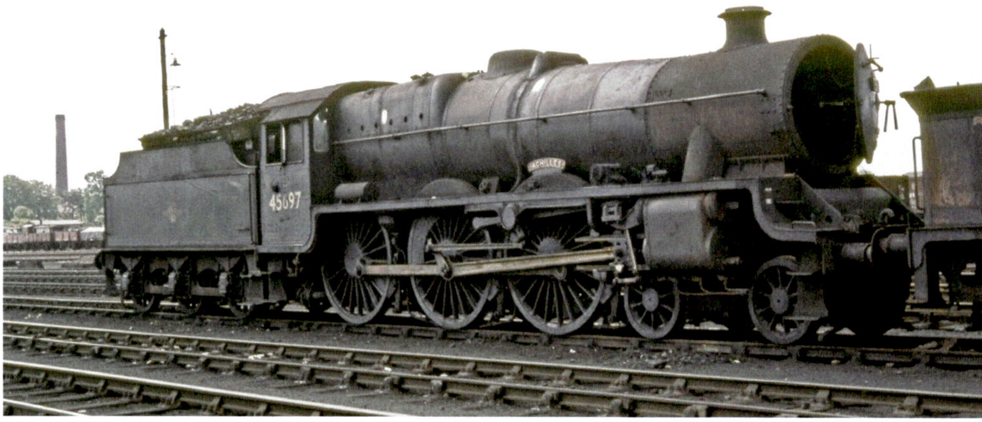

24 June 1962 LMS Jubilee 4-6-0 45697 at Carlisle Kingmoor MPD.
LMS Jubilee 4-6-0 45697 *Achilles* was a product of Crewe Works in April 1936. It had been a resident of Carlisle Kingmoor for ten years until September 1962 when it moved on to Blackpool Central MPD. In March 1964 it was transferred to Leeds Holbeck and became one of the last three Jubilees in service in the summer of 1967. After withdrawal in September of that year, it donated many of its parts to 45593 before the remains were cut up by Cashmore's at Great Bridge. It was seen at Carlisle paired with a Fowler tender which was replaced by Stanier pattern tender at its final works visit soon after this photograph was taken.

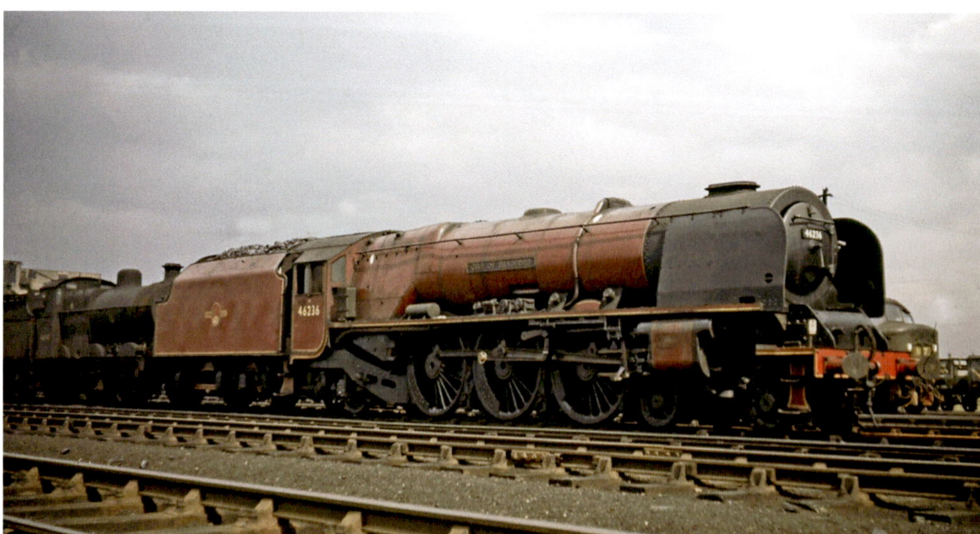

April 1961 LMS Coronation 4-6-2 46236 at Carlisle Kingmoor MPD.
LMS Coronation 4-6-2 46236 *City of Bradford* was a Crewe product dating from July 1939 and had spent much of its career based at Camden and Crewe. In March 1961 it was reallocated to Carlisle Upperby and soon after to Kingmoor where it remained active until withdrawal in March 1964. It did not receive a yellow diagonal stripe on its cab because these were applied in the summer of 1964.

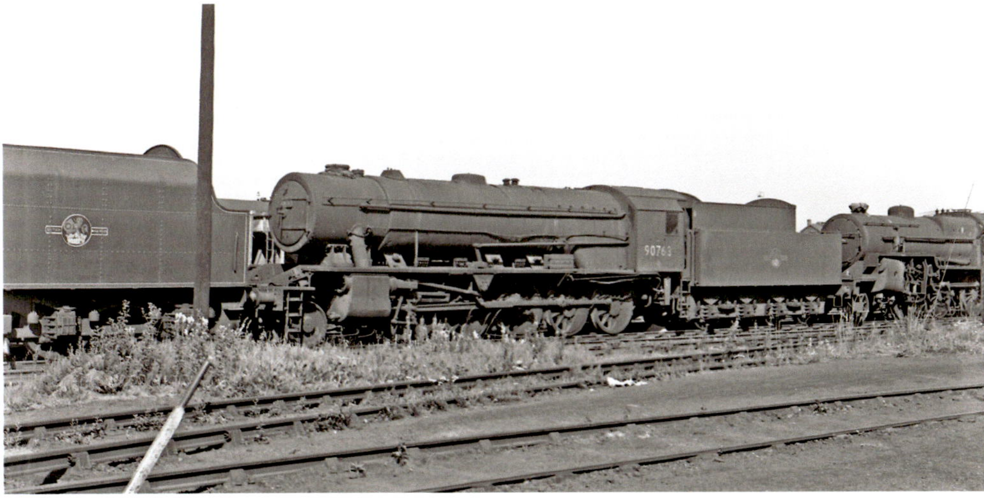

August 1963 WD 2-10-0 90763 at Carlisle Kingmoor MPD.
WD 2-10-0 90763 had been built by the North British Locomotive Company of Glasgow for the Ministry of Supply in June 1945 as 73787. It was immediately loaned to the LNER and worked from March MPD until recalled by the War Department in October 1946. After storage at the Longmoor Military Railway it became part of the British Railways fleet of 2-10-0s in December 1948. It was unusual among the other WD 2-10-0s because it spent much of its career in England, mainly based at Carlisle Kingmoor, but also allocated for twelve months to Bidston, Birkenhead and a few weeks at Doncaster. It was withdrawn along with the remaining Scottish WD 2-10-0s in December 1962 and was cut up at Darlington.

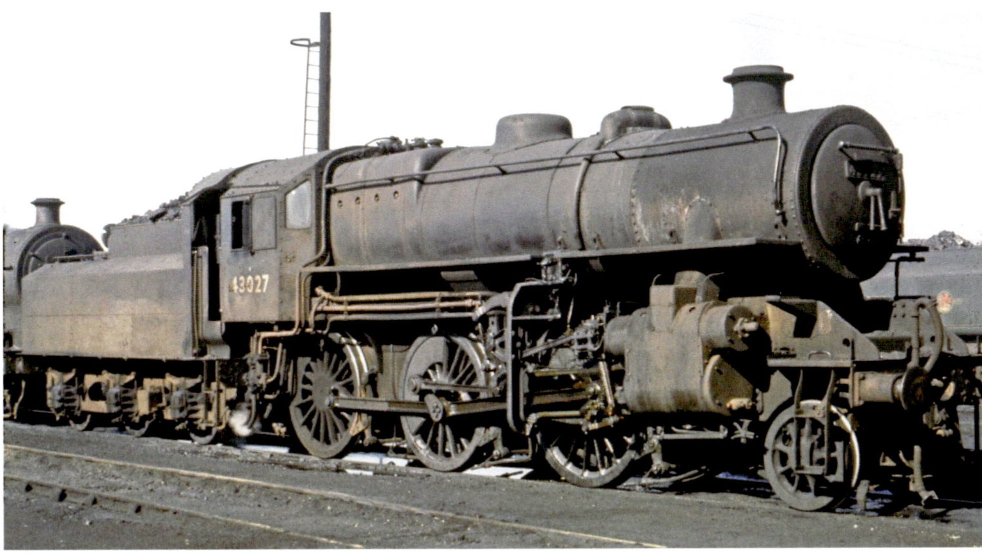

18 April 1964 LMS Ivatt 4MT 2-6-0 43027 at Carlisle Kingmoor.
LMS Ivatt 4MT 2-6-0 43027 was built for British Railways by Horwich Works in February 1949 and spent ten years based at depots in the Midlands. It was transferred to Carlisle Kingmoor in July 1960 and then on to Carnforth in October 1964. In January 1968 it moved for the last time to Lostock Hall MPD, but it was to be a short stay because withdrawal came a few months later in May, and by September it had been cut up by Arnott Young of Dinsdale.

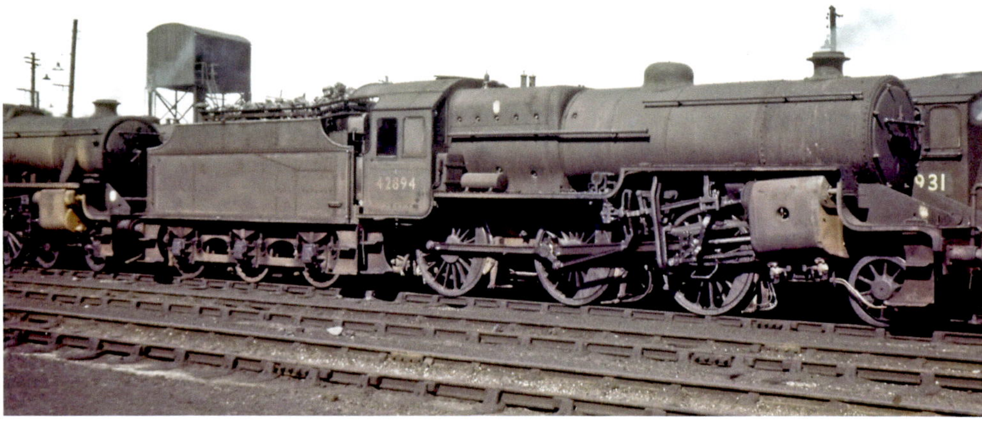

April 1964 LMS Crab 5P6F 2-6-0 42894 at Carlisle Kingmoor MPD.
LMS Crab 5P6F 2-6-0 42894 had been built at Crewe Works in August 1930. It had been allocated to depots in the Midlands and the North West throughout its active life. When it was seen at Carlisle Kingmoor it was allocated to Springs Branch, Wigan, but would spend its last four months at Birkenhead Mollington Street. It was withdrawn from there in December 1964 and cut up by T. W. Ward of Sheffield in March 1965.

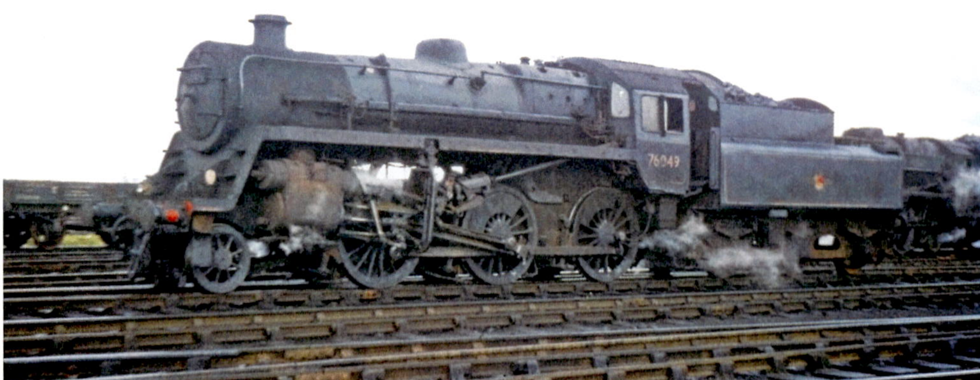

April 1964 BR Standard 4MT 2-6-0 76049 at Carlisle Kingmoor MPD.
BR standard 4MT 2-6-0 76049 had been built at Doncaster Plant in April 1955. It spent the first eight years of its life allocated to depots in the North East. It was then transferred to Scotland, in January 1963, to Hawick and then to St Margaret's in October 1965. It was withdrawn from Bathgate in January 1966 days after transfer to that depot and cut up by Motherwell Machinery and Scrap at Wishaw in May 1966.

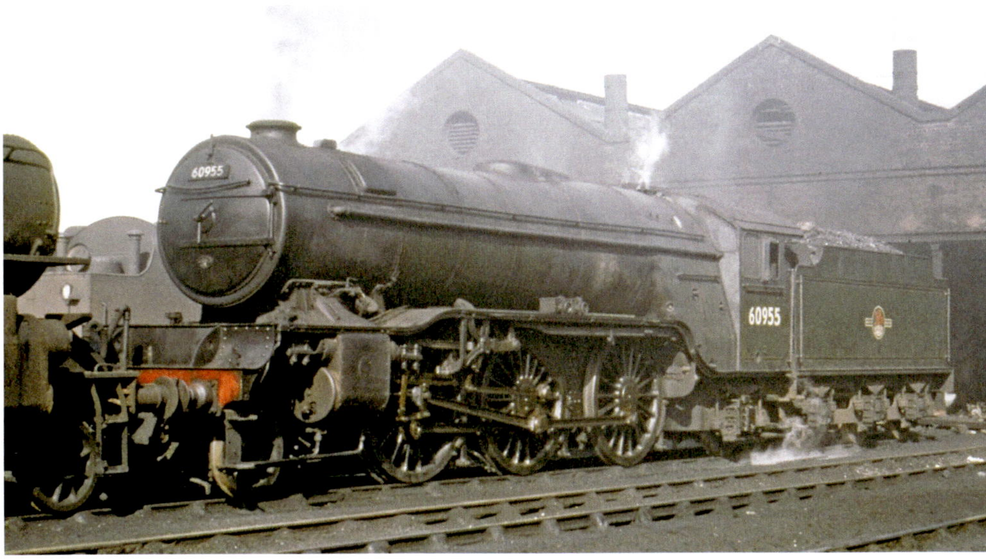

May 1964 LNER 4-6-2 V2 60955 at Carlisle Kingmoor MPD.
LNER 4-6-2 V2 60955 had been built at Darlington Works in September 1942 and had been based in Scotland throughout its working life. Its first allocation was to Haymarket MPD. At the formation of British Railways it was transferred to Aberdeen Ferryhill. Its final home was St Margaret's, where it moved to in May 1964, and was withdrawn from there in September 1966. It was one of the final fourteen survivors of the class; all but one had succumbed to the cutter's torch by the end of 1966. (Photograph L. Flint)

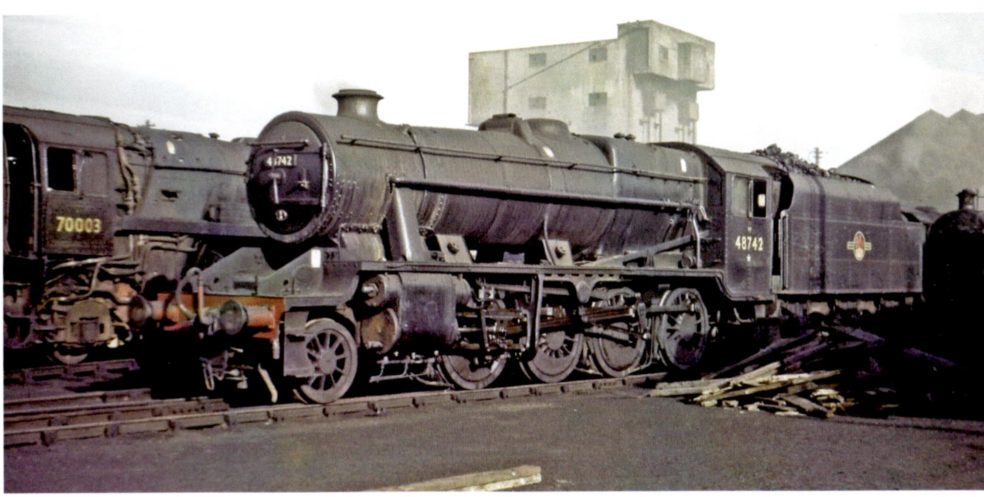

June 1964 LMS 8F 2-8-0 48742 at Carlisle Kingmoor.
LMS 8F 2-8-0 48742 had been built for the LNER at Darlington Works in March 1946 as a Class O6, one of twenty-three 8Fs built there. The LNER owned sixty-eight of the class at the end of the Second World War, but they had been sold to the LMS before the formation of British Railways in January 1948. This locomotive began its BR career based at Buxton MPD before moving on to Northwich, then in May 1960 to Edge Hill. When photographed at Kingmoor it was still looking smart and reasonably clean after its general overhaul six weeks earlier. It was withdrawn from Newton Heath in September 1967 and cut up by Cashmore's of Great Bridge in January 1968.

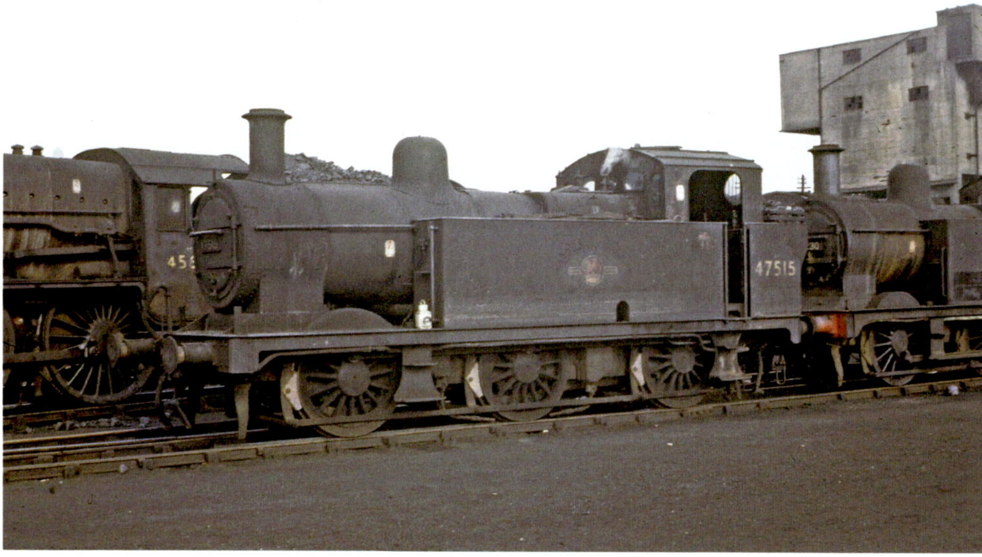

June 1964 LMS Jinty 3F 0-6-0T 47515 at Carlisle Kingmoor MPD.
LMS Jinty 3F 0-6-0T 47515 was built by Vulcan Foundry for the LMS in May 1928. For many years it had been at Devon's Road MPD, Bow. It was reallocated to Carlisle Kingmoor in September 1958 – it was a few weeks away from withdrawal when seen here in steam on shed. It was cut up by Campbell's of Airdrie in January 1965.

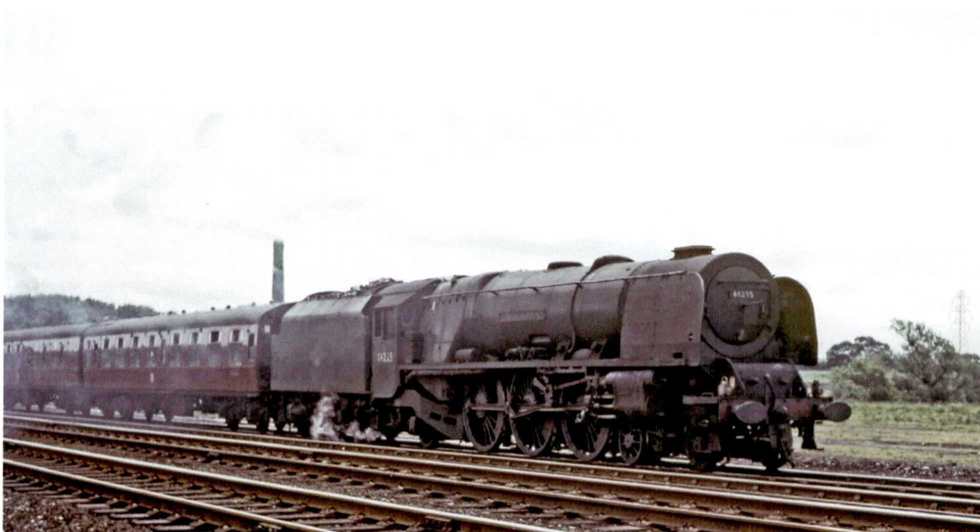

July 1964 LMS Coronation 4-6-2 46255 passes Carlisle Kingmoor.
LMS Coronation 4-6-2 46255 *City of Hereford* had been built at Crewe Works in October 1946 and had been allocated to Carlisle Upperby three times in its career. The last time it moved to Upperby was in February 1956 and it eventually moved across the city in April 1961 to Carlisle Kingmoor. It was to remain active until the mass culling of the class in September 1964 when the survivors were all withdrawn. It was cut up by Arnott Young of Troon in early 1965.

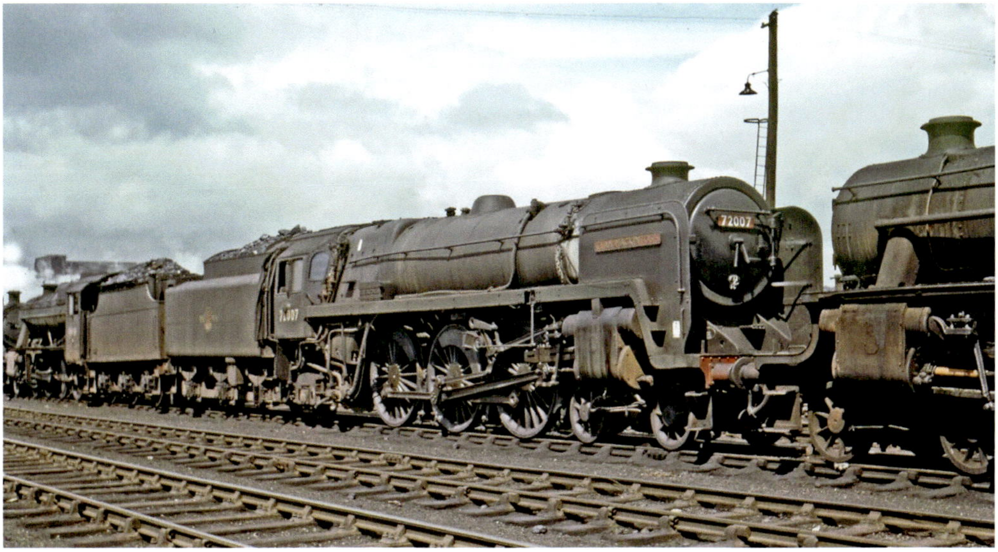

August 1964 BR Standard Clan 6MT 4-6-2 72007 at Carlisle Kingmoor MPD.
BR Standard Clan 6MT 4-6-2 72007 *Clan Mackintosh* had been built at Crewe Works in February 1952; it had always been allocated to Carlisle Kingmoor. It had been used in May 1964 for the RCTS Ribble-Lune Rail Tour when it looked in immaculate condition – three months later the paintwork was gradually disappearing under a coat of grime. It was withdrawn in December 1965 and cut up by Campbell's of Airdrie in March 1966.

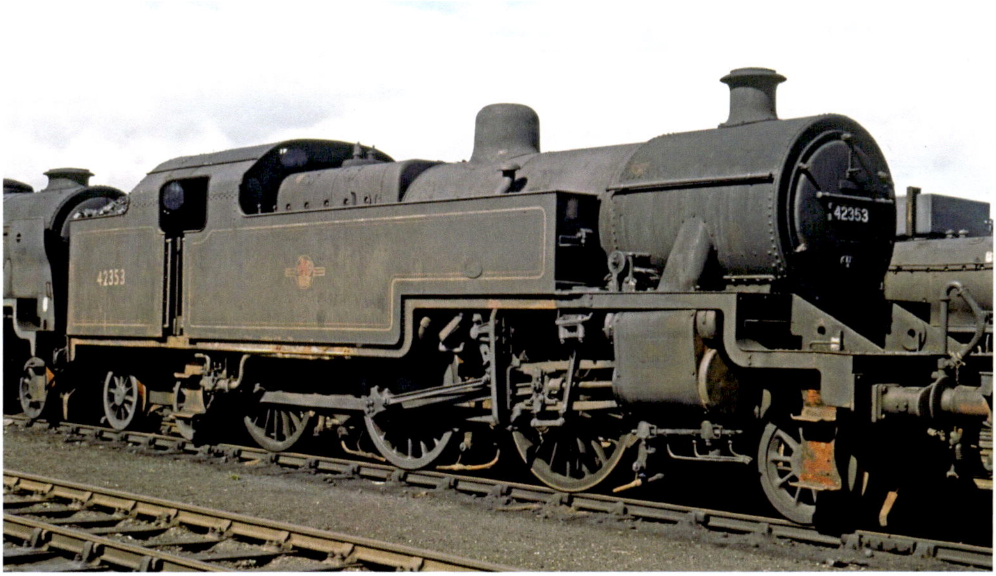

12 July 1964 LMS Fowler 4MT 2-6-4T 42353 at Carlisle Kingmoor MPD.
LMS Fowler 4MT 2-6-4T 42353 had been built at Derby Works in June 1929. It was allocated to several depots in the North West and North Wales before reallocation to Carlisle Kingmoor in November 1963. There was little work for the locomotive by this time and although fully coaled when seen here, it had already been withdrawn in June 1964. By September it was cut up by Arnott Young of Carmyle.

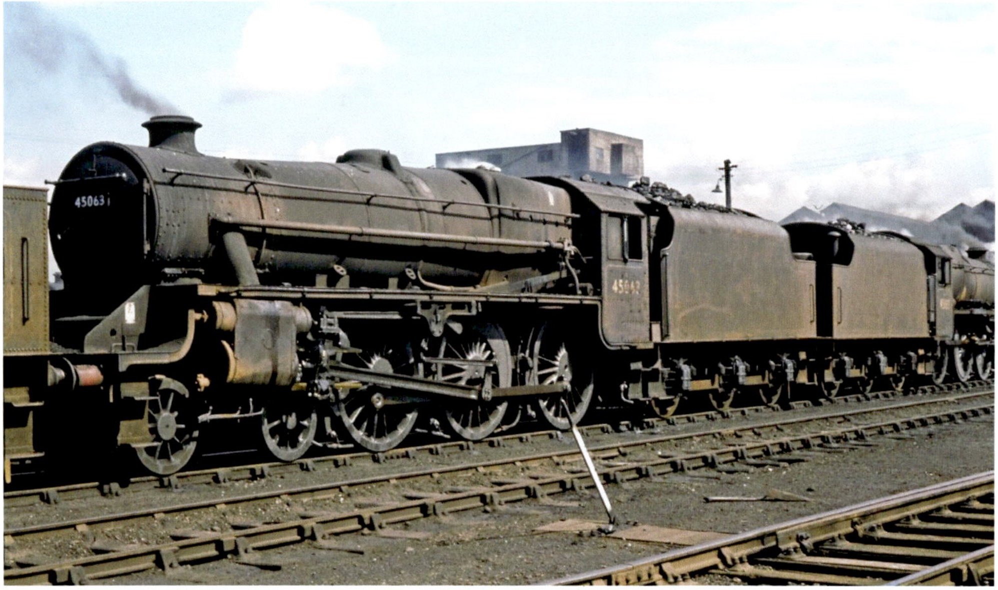

12 July 1964 LMS Black Five 4-6-0 45063 at Carlisle Kingmoor.
LMS Black Five 4-6-0 45063 had been built by Vulcan Foundry for the LMS in December 1934. It had spent most of its active life working from West Yorkshire, for many years based at Farnley Junction MPD, and from June 1964 at Holbeck depot. It was withdrawn from service in October 1966 and cut up by Draper's of Hull in January 1967. This locomotive was part of the first batch of fifty Black Fives to be built by a private manufacturer and was reunited with a dome-less boiler for its last years of operation.

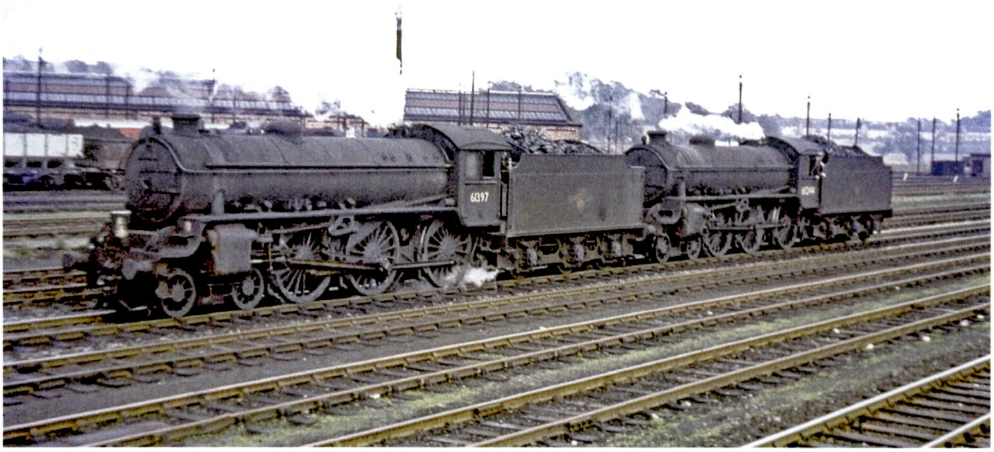

October 1964 LNER B1 4-6-0s 61397 and 61244 near Carlisle Kingmoor.
LNER B1 4-6-0s 61397 and 61244 *Strang Steel* back down past Carlisle Kingmoor depot. Both locomotives had been built by the North British Locomotive Company of Glasgow – 61397 for British Railways in March 1952 and 61244 for the LNER in October 1947. They had been allocated to depots in Scotland throughout their working lives; 61244 was reallocated to St Margaret's MPD in October 1961, while 61397 had been at that depot from new. Both locomotives were withdrawn in 1965 and both were cut up by Shipbreaking Industries of Faslane.

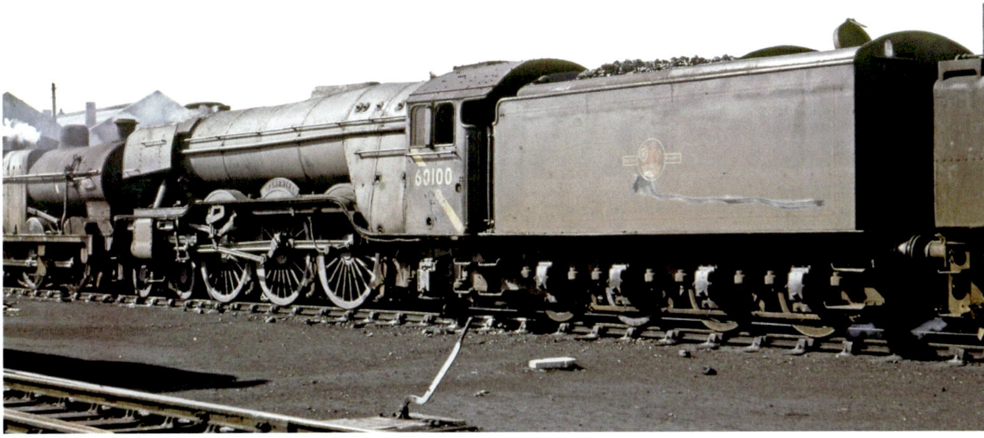

November 1964 LNER A3 4-6-2 60100 at Carlisle Kingmoor MPD.
LNER A3 4-6-2 60100 *Spearmint* had been built at Doncaster Plant and entered service in May 1930. It had always been allocated to depots in Scotland and spent thirty of its thirty-five years' active service based at Haymarket MPD. In January 1963 it was reallocated to St Margaret's depot and became a familiar sight on the Waverley Route. In August 1964 it became one of only three A3s to carry the diagonal cab-side stripe which denoted the locomotive was forbidden to work south of Crewe, under the then new overhead electric wires. It was one of a trio of A3s to be still active into 1965, but was the first to be withdrawn in June of that year, when it visited Darlington Works for repair. It was found to have a severely cracked frame and was cut up at the works soon afterwards.

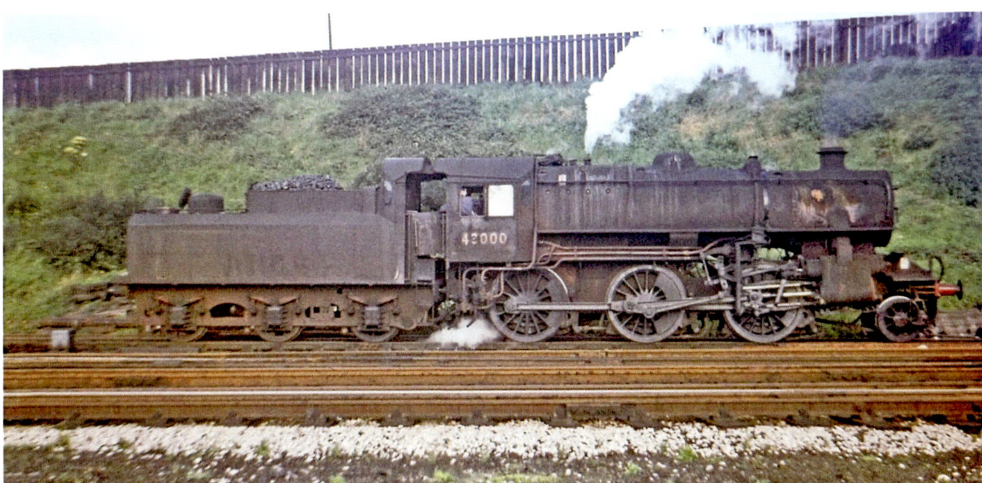

May 1965 LMS Ivatt 4MT 2-6-0 43000 near Carlisle Kingmoor MPD.
LMS Ivatt 4MT 2-6-0 43000 had been built at Horwich Works for the LMS in December 1947, being one of three completed that year. By the time the locomotive had entered service it was under the ownership of the newly formed British Railways. It spent its first nine years in the London area, mainly working from Devons Road, Bow depot, before reallocation to Nuneaton, where it stayed for three years. In April 1961 it moved north to Carlisle Upperby and then spent six months at Carlisle Canal before arriving at Carlisle Kingmoor in June 1963. One final reallocation took the engine to Blyth North MPD from where it was withdrawn in September 1967. It was scrapped by Clayton and Davie of Dunston-on-Tyne in April 1968.

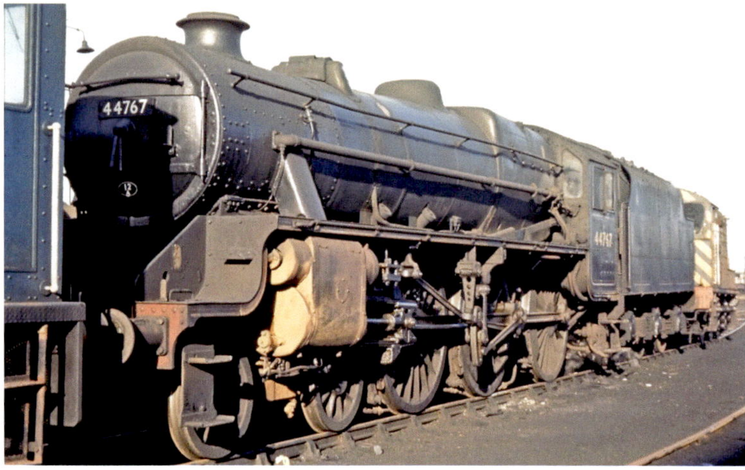

May 1965 LMS Black Five 4-6-0 44767 at Carlisle Kingmoor MPD.
LMS Black Five 4-6-0 44767 was built at Crewe Works and completed on the last day of the London Midland and Scottish Railway in December 1947. The year 1948 brought with it the beginning of the nationalised British Railways. 44767 was unique among the Black Fives, being fitted with Stephenson link motion as part of an experiment carried out by George Ivatt, the Chief Mechanical Engineer of the LMS. He hoped to improve an already successful design. The locomotive was allocated to Bank Hall MPD for twelve years before moving on to Southport and finally to Carlisle Kingmoor in November 1964. After withdrawal in December 1967, it was purchased for preservation and returned to steam in 1975. Since that time, it has worked regularly on the main line and on heritage railways. (Photograph L. Flint)

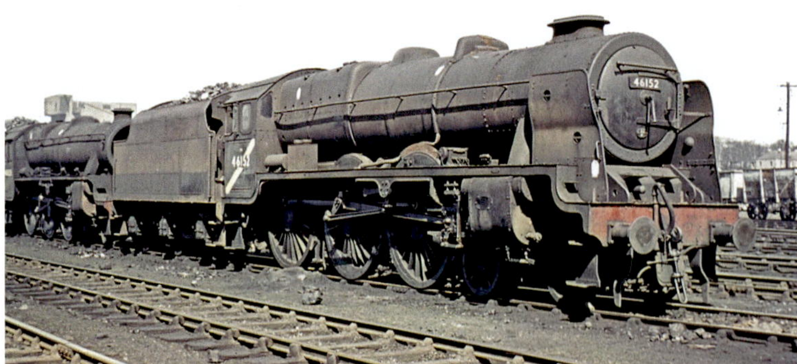

May 1965 LMS Royal Scot 4-6-0 46152 at Carlisle Kingmoor MPD.
LMS Royal Scot 4-6-0 46152 *The King's Dragoon Guardsman* was built for the LMS by the North British Locomotive Company of Glasgow in August 1927. It was rebuilt with a tapered boiler, new cylinders and frames in 1945. This locomotive had been built as 6100 Royal Scot but changed identities with the original 6152 in 1933. At the time 6152 was in better condition and was chosen to represent the LMS at the Chicago Century of Progress exposition. 46152 was one of the last three survivors of the class which all worked from Carlisle Kingmoor into 1965. When seen there in May 1965, it had been withdrawn a short time before but was still fully coaled. By the end of the summer, it was to be cut up by Motherwell Machinery and Scrap of Wishaw.

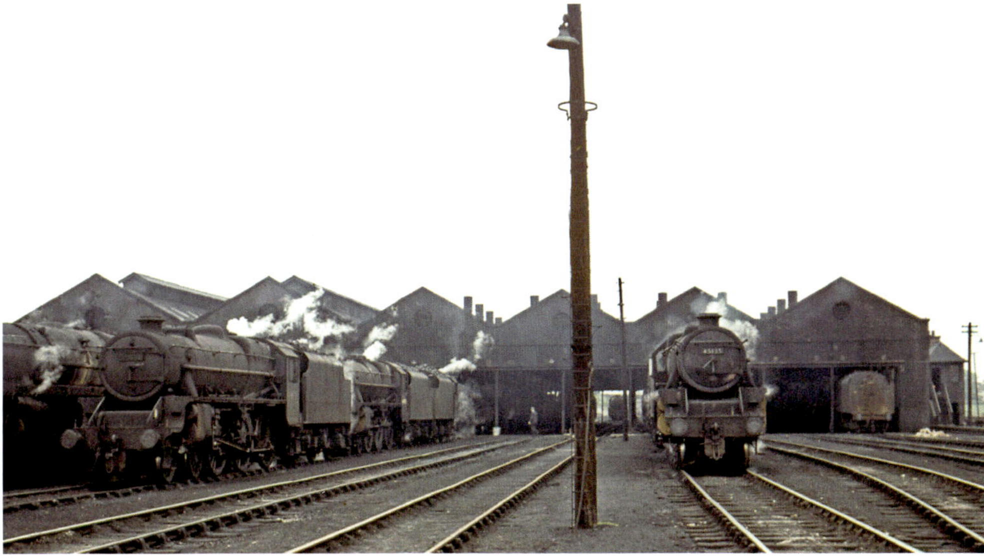

16 May 1965 LMS Black Five 4-6-0s including 45135 at Carlisle Kingmoor MPD.
LMS Black Five 4-6-0s including 45135 in steam at Carlisle Kingmoor await their next duty. 45135 had been built for the LMS by Armstrong Whitworth of Newcastle and had entered service in May 1935. During its British Railways career it had been allocated to six different depots, with its longest stay of ten years at Springs Branch, Wigan. It became a Kingmoor loco in November 1964 and worked from there until its withdrawal in October 1967. It was cut up by J. McWilliam of Shettleston in April 1968.

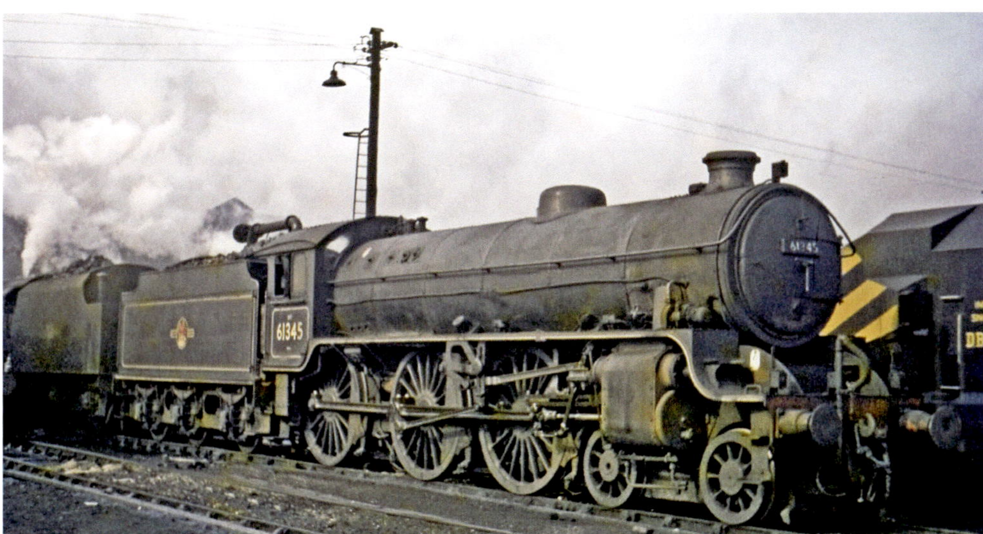

June 1965 LNER B1 4-6-0 61345 at Carlisle Kingmoor MPD.
LNER B1 4-6-0 61345 was built by Gorton Works for British Railways and entered service on the Scottish Region in April 1949. It worked from Kittybrewster depot until June 1961 when it was transferred to St Margaret's MPD. The curved 'infills' on the running plate corners, which were applied to Scottish B1s, are clearly visible. The locomotive was withdrawn in July 1966 and was cut up by G. H. Campbell of Airdrie in October 1966.

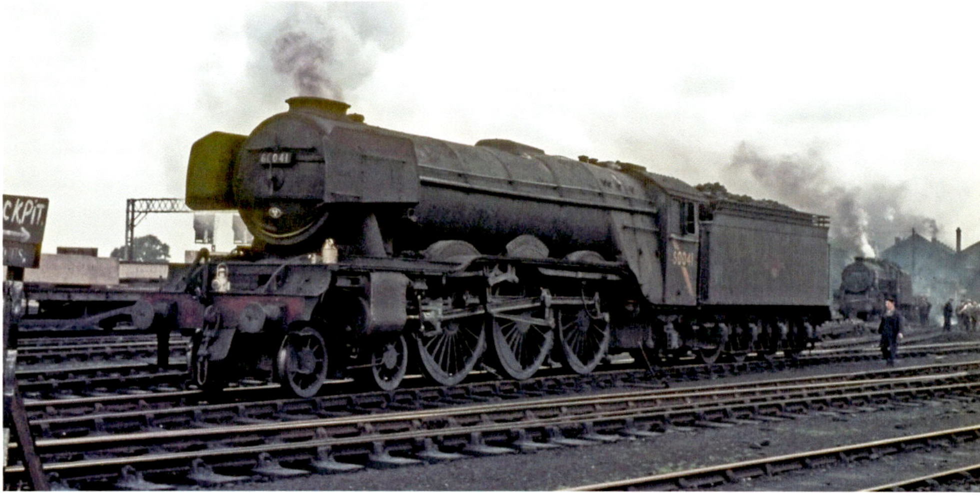

August 1965 LNER A3 4-6-2 60041 at Carlisle Kingmoor MPD.
LNER A3 4-6-2 60041 *Salmon Trout* had been built at Doncaster Works in December 1934 and had spent its working life allocated to Scottish depots, mainly at Haymarket and then in July 1960 at St Margaret's. Its last general overhaul was completed at Doncaster in January 1963 when it was also fitted with trough-type smoke deflectors. When withdrawn in December 1965 it had become the penultimate A3 in service. It was stored at Cowlairs Works until the summer of 1966 with some parts removed for use with the preserved A3 *Flying Scotsman* including its boiler and cylinders. The remains were cut up by Arnott Young of Carmyle in September 1966. (Photograph L. Flint)

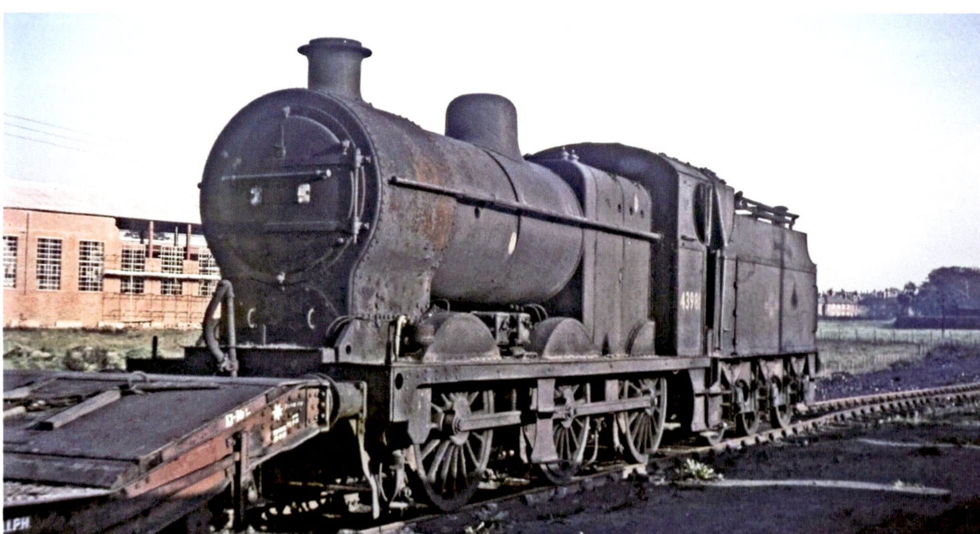

August 1965 MR 4F 0-6-0 43981 out of use at Carlisle Kingmoor MPD.
MR 4F 0-6-0 43981 had been built by Armstrong Whitworth of Newcastle for the Midland Railway in 1922. It had begun its British Railways career allocated to Spital Bridge MPD, Peterborough, on the Eastern Region. It later moved to several different depots in Wales before its final reallocation to Carlisle Kingmoor in October 1962. It was seen here withdrawn from Kingmoor in August 1965. It was cut up by Bird's at Tremain's Yard in Bridgend.

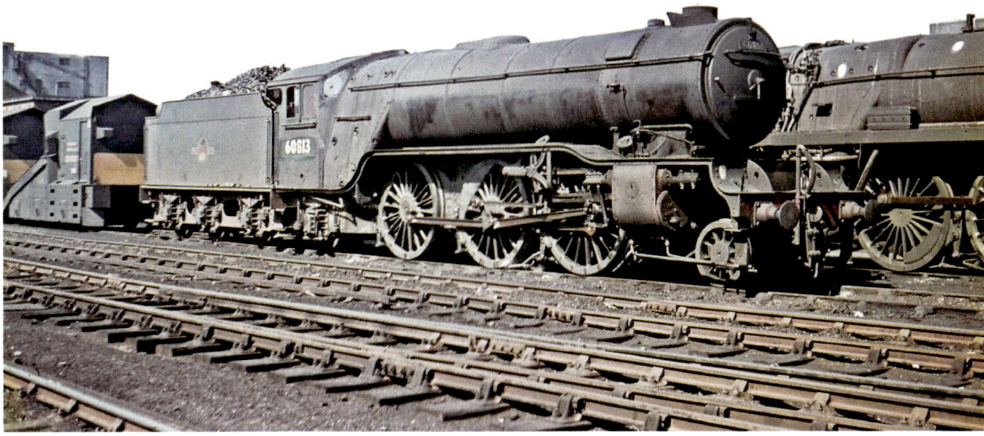

August 1965 LNER 2-6-2 V2 60813 at Carlisle Kingmoor MPD.
LNER 2-6-2 V2 60813 had been built at Darlington Works in September 1937. The locomotive was unique among other V2s in having a stovepipe chimney and a wrap-around smoke deflector which was fitted in 1946. It retained these for the rest of its career. It was allocated to Doncaster and King's Cross depots for its first fourteen years of service. In February 1951 it moved to Scotland, first to St Margaret's and finally in May 1966 to Dundee Tay Bridge. In September 1966 it was withdrawn and was cut up by J. McWilliam of Shettleston.

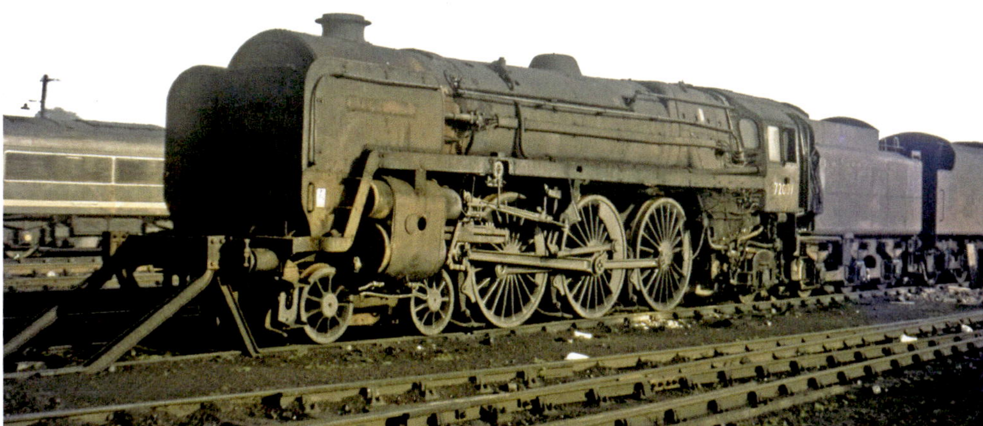

August 1965 BR Standard Clan 6MT 4-6-2 72009 at Carlisle Kingmoor MPD.
BR Standard Clan 6MT 4-6-2 72009 *Clan Stewart* had been allocated to Carlisle Kingmoor since it emerged from Crewe Works in March 1952. In 1958 it strayed away from Kingmoor for a short time when it was loaned to Stratford depot for a few weeks to assess its suitability of working over the Great Eastern lines to Liverpool Street station. It was placed in store in the summer of 1965 and had its tender emptied of coal. It was officially withdrawn in November 1965 and cut up by the Motherwell Machinery and Scrap of Wishaw.

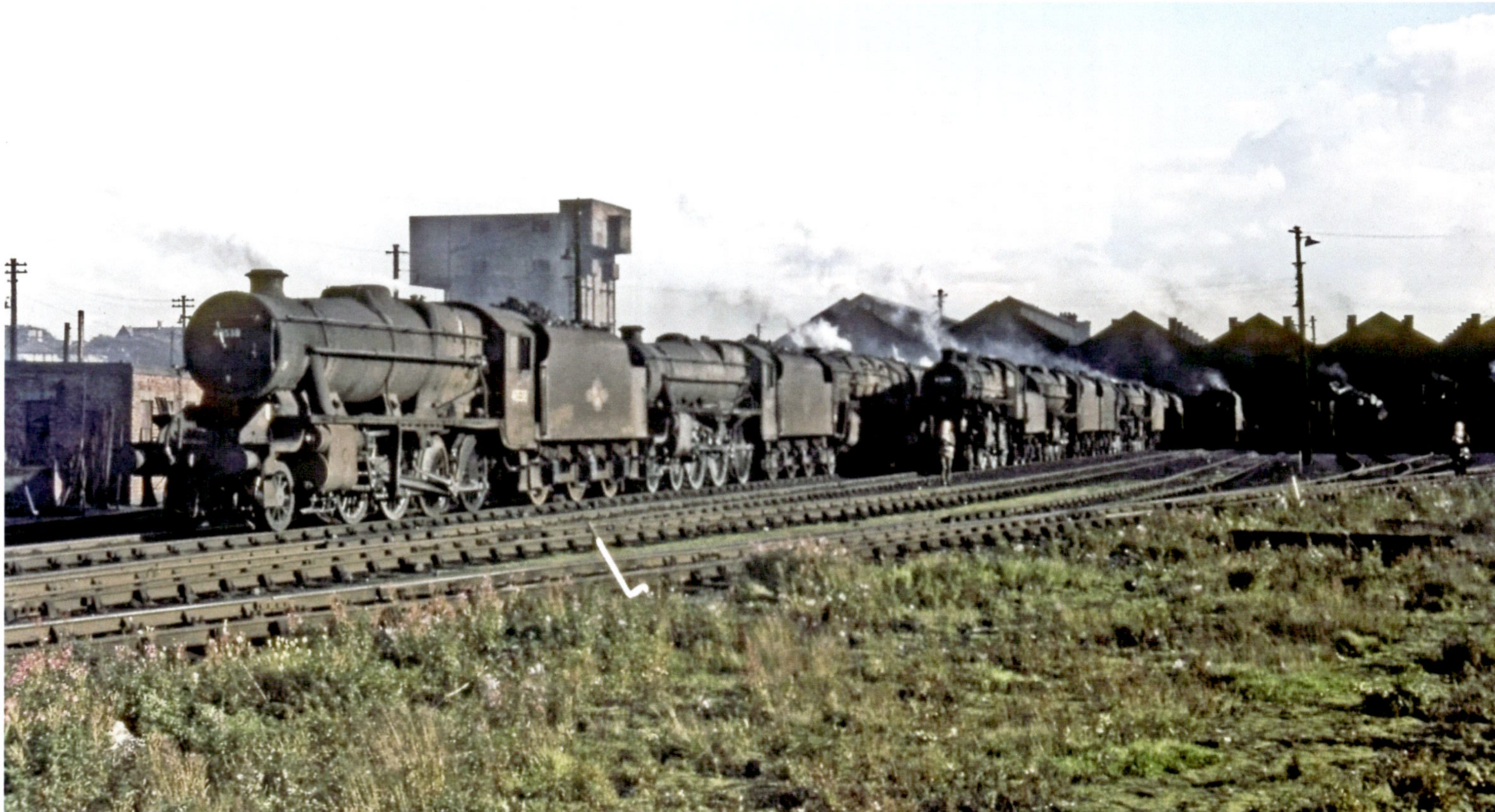

September 1965 LMS 8F 2-8-0 48538 at Carlisle Kingmoor MPD.
LMS 8F 2-8-0 48538 was at the head of a line of locomotives fully coaled, in steam and awaiting their next duty. 48538 had been built at Doncaster Plant in August 1945 and had spent most of its working life allocated to depots in the East Midlands. It was at Toton for nine years and was also allocated for spells at Wellingborough, Colwick and Westhouses. It was a Westhouses MPD locomotive when photographed at Carlisle Kingmoor, but was to end its days at Saltley depot. It was withdrawn in March 1967 and was taken to Cashmore's of Great Bridge for scrapping in the summer of that year.

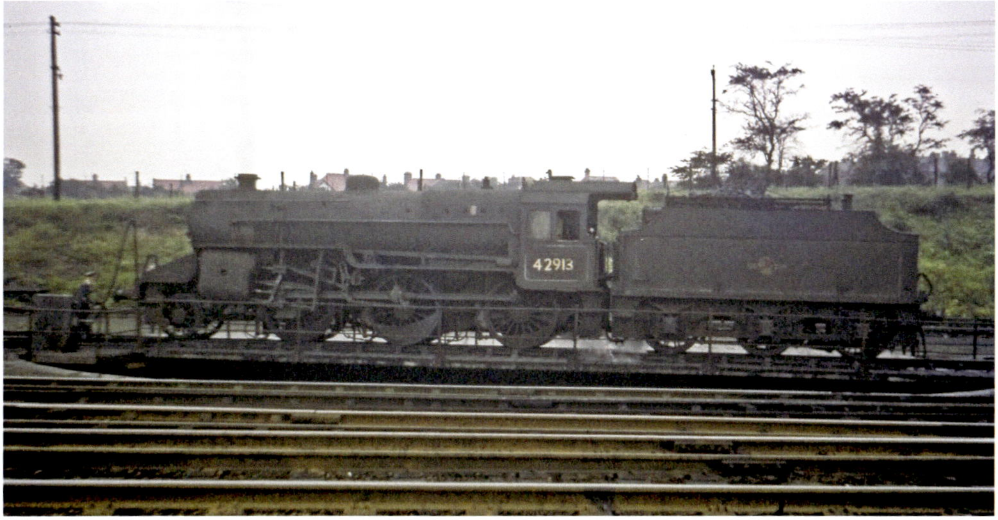

November 1965 LMS Crab 5P6F 2-6-0 42913 on the turntable at Carlisle Kingmoor.
LMS Crab 5P6F 2-6-0 42913 was seen on the turntable at Carlisle Kingmoor as it was being prepared for its return working to its home shed of Ayr. It had been built at Crewe Works in October 1930 and at the dawn of British Railways it was already in Scotland, allocated to Corkerhill. Its only reallocation came in May 1964 when it joined other members of the class at Ayr MPD. It was finally withdrawn from there in June 1966 and scrapped by Arnott Young of Troon in October 1966.

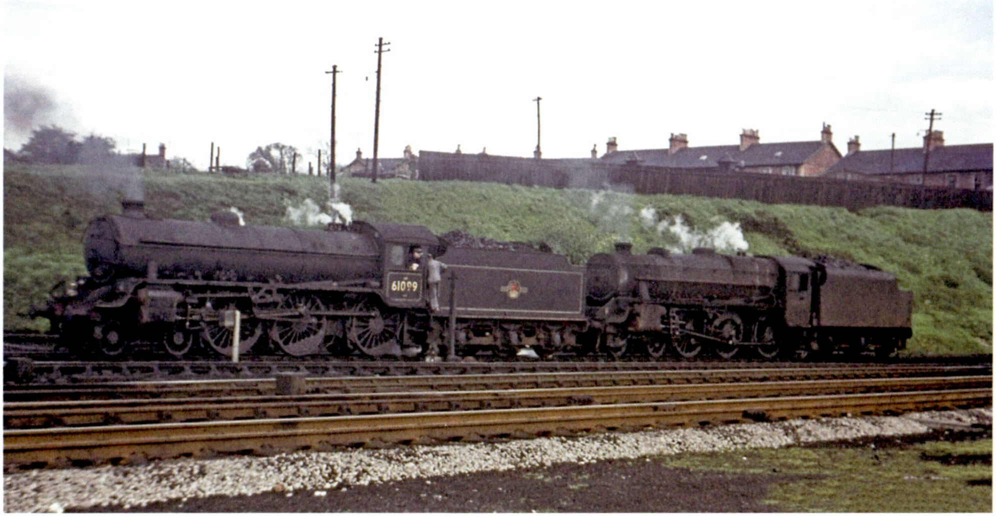

November 1965 LNER B1 4-6-0 61099 with an LMS Black Five near Carlisle Kingmoor.
LNER B1 4-6-0 61099 was coupled together with an unidentified LMS Black Five as they prepared to leave the environs of Carlisle Kingmoor depot. 61099 had been built by the North British Locomotive Company of Glasgow, for the LNER in November 1946, and had spent the first five years working in the London area, particularly from Hitchen MPD. In September 1953 it was transferred to St Margaret's depot in Edinburgh and remained active from there until its final move in April 1966 to Thornton Junction MPD. Within a few months it had been withdrawn from there and was cut up by Shipbreaking Industries of Faslane by the end of 1966.

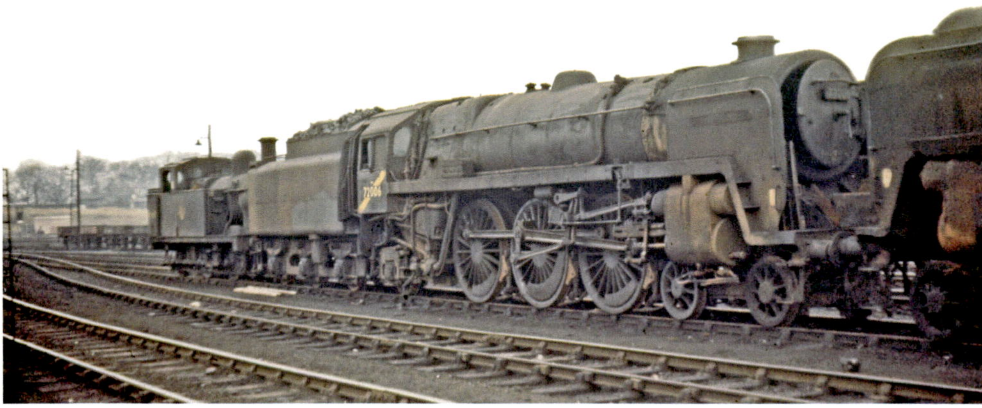

February 1966 BR Standard Clan 6MT 4-6-2 72006 at Carlisle Kingmoor MPD.
BR Standard Clan 6MT 4-6-2 72006 *Clan Mackenzie* was still in service when photographed on Kingmoor depot. It looked very neglected but was fully coaled and appeared to be having work carried out in the smokebox. In the summer of 1964, it had been wrongly painted with a diagonal stripe, which prohibited it from working south of Crewe under the electric overheads. In May 1966 it was withdrawn from service.

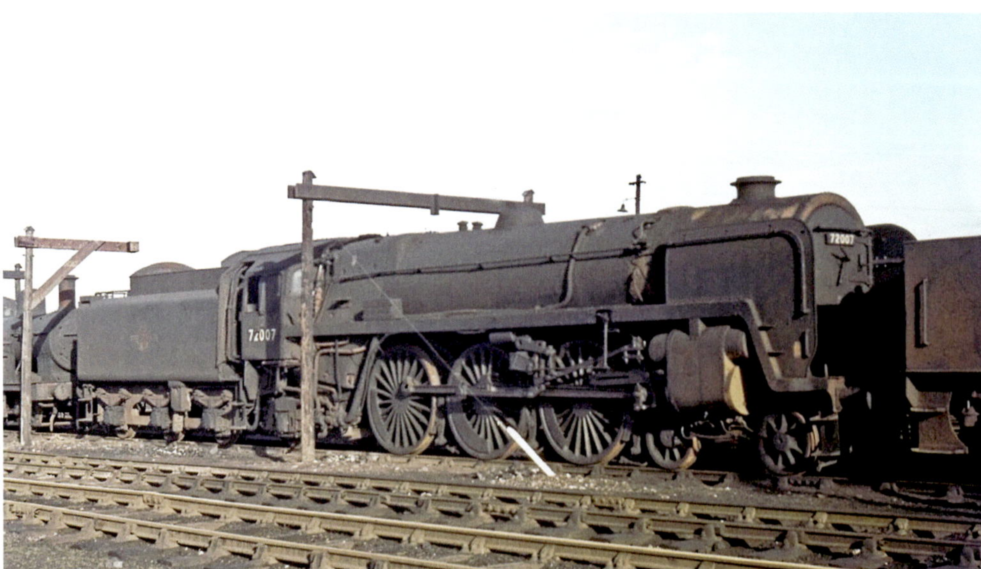

February 1966 BR Standard Clan 6MT 4-6-2 72007 at Carlisle Kingmoor MPD.
BR Standard Clan 6MT 4-6-2 72007 *Clan Mackintosh* was the only other member of the class to remain in service until 1966, although, as can been seen, it certainly was not in active service. It waited for its final call to the scrapyard with a line of other withdrawn locomotives. It was officially withdrawn in April and was taken to Campbell's yard in Airdrie for scrapping in September 1966.

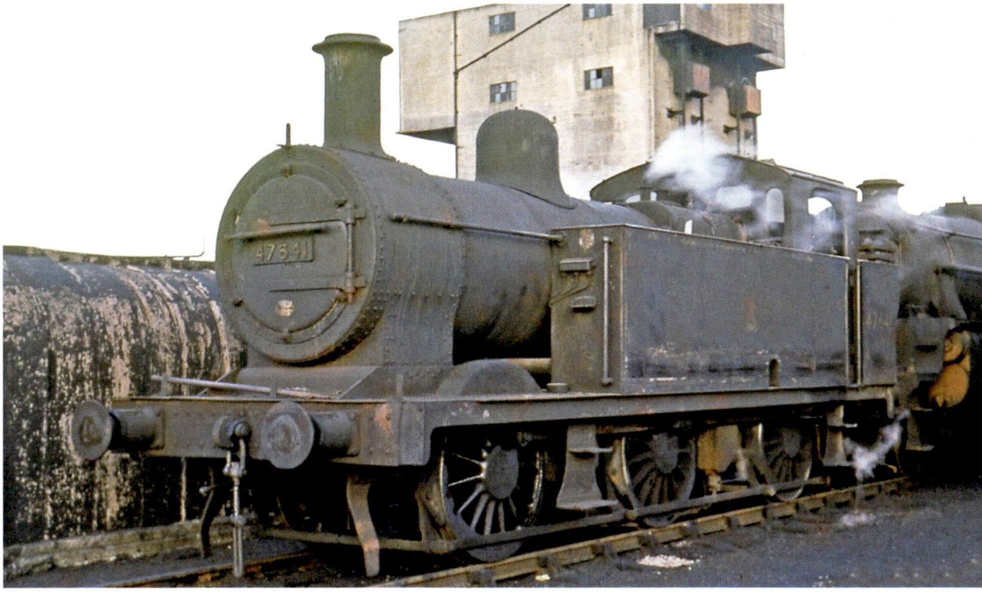

April 1966 LMS Jinty 3F 0-6-0T 47641 at Carlisle Kingmoor MPD.
LMS Jinty 3F 0-6-0T 47641 was built by Beardmore of Dalmuir, Glasgow, in January 1929 for the LMS. It spent the first fourteen years of its British Railways career allocated to Burton MPD, then two more years at Gorton before finding a home at Carlisle Kingmoor in July 1964. It was finally withdrawn in December 1966 but remained in store at Kingmoor until the closure of the depot to steam a year later. In February 1968 it had found its way to J. McWilliam in Shettleston where it was cut up.

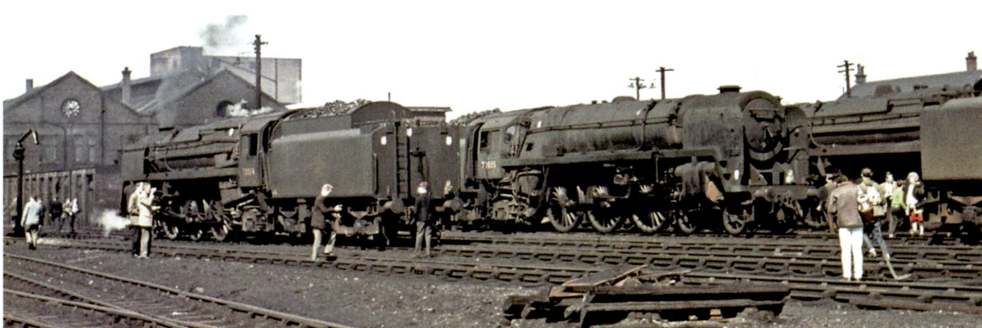

30 April 1966 BR Standard Britannia 7MT 4-6-2s 70014 and 70035 at Carlisle Kingmoor MPD.
BR Standard Britannia 7MT 4-6-2s 70014 *Iron Duke* and 70035 *Rudyard Kipling* were part of the great gathering of the Britannia class at Carlisle Kingmoor in the last years of steam. Both locomotives had very different careers before their reallocation to Carlisle. 70014 had been built at Crewe in June 1951 and had spent its first seven years allocated to the Southern Region at Nine Elms and then to Stewarts Lane, where it was responsible for hauling the prestigious 'Golden Arrow' train along with classmate 70004. In July 1958 the Britannias were exchanged by the Southern Region for two Standard 5 4-6-0s and were reallocated to Kentish Town. A nomadic life followed for 70014 until its final reallocation to Kingmoor in the summer of 1966 where it stayed along with 70035 until the depot's closure in December 1967. After a few months in store the pair were taken to Inverkeithing and cut up by T. W. Ward in the spring of 1968. (Photograph L. Flint)

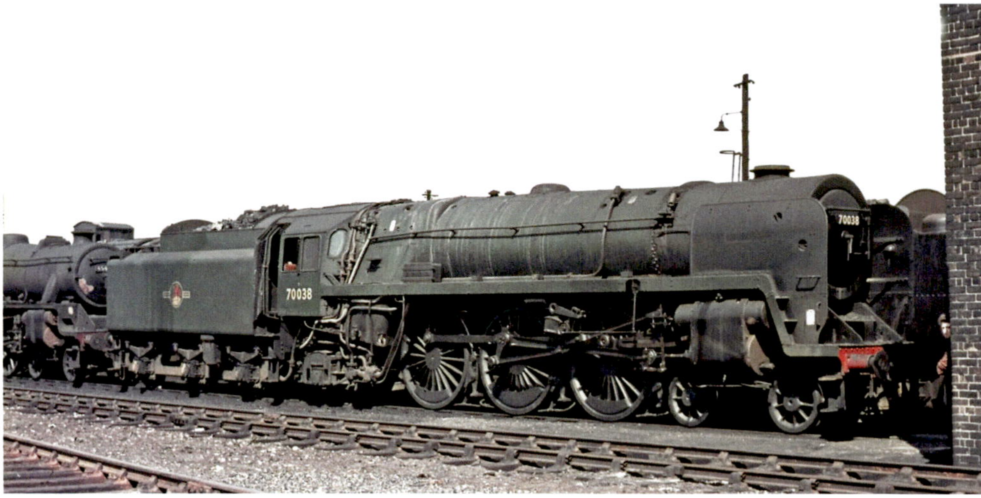

30 April 1966 BR Standard Britannia 7MT 4-6-2 70038 at Carlisle Kingmoor MPD.
BR Standard Britannia 7MT 4-6-2 70038 *Robin Hood* had been built at Crewe Works in January 1953. It was another of the Great Eastern-based Britannias which were transferred to Immingham depot in October 1961. Their reallocation and two-year stay at Immingham was amazing to me, at the time a young trainspotter living close to the Cleethorpes to Doncaster line. By December 1963 they had been transferred away to Carlisle Kingmoor. 70038 had returned from Crewe Works a few weeks before this photograph, but already its unlined green livery was looking shabby. It was withdrawn in August 1967 and was cut up by J. McWilliam of Shettleston in early 1968. (Photograph L. Flint)

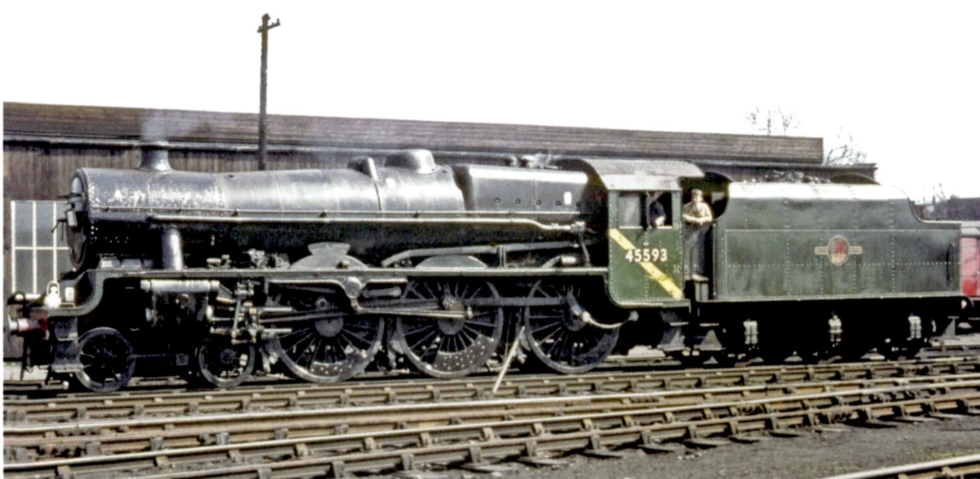

30 April 1966 LMS Stanier Jubilee 4-6-0 45593 at Carlisle Kingmoor.
LMS Stanier Jubilee 4-6-0 45593 was into its last eighteen months of service when it was photographed at Carlisle Kingmoor. It had worked The Jubilee Railway Society's rail tour named 'The South Yorkshireman No. 5' from Bradford Foster Square via Carnforth and the West Coast Main Line to Carlisle. Although looking externally in fine fettle as it glinted in the spring sunshine, the locomotive was actually failed at Kingmoor and the return journey of the special was hauled by BR Standard Britannia 7MT 4-6-2 70035 *Rudyard Kipling*.

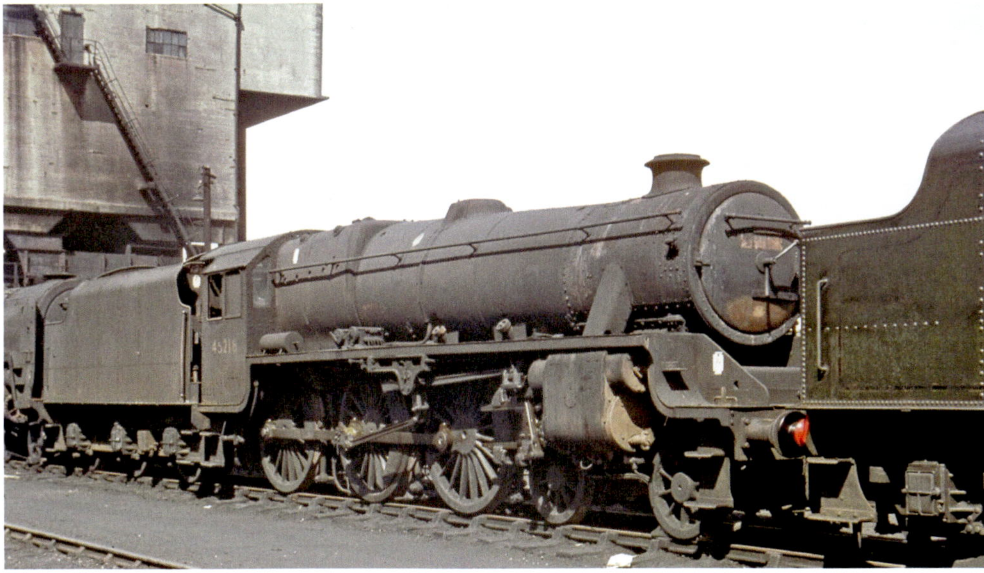

May 1966 LMS Black Five 4-6-0 45218 at Carlisle Kingmoor MPD.
LMS Black Five 4-6-0 45218 had been built by Armstrong Whitworth of Newcastle for the LMS in November 1935. It had been allocated to depots around the North West before spending eight years in West Yorkshire at the beginning of British Railways in January 1948. After another move and an eight-year stay at Southport MPD, it was reallocated to Carlisle Canal in January 1963. With the closure of Canal depot in June 1963 it moved across to Kingmoor for six months before heading back to various depots in the North West. In September 1965 it returned to Carlisle Kingmoor and was withdrawn from there in April 1966. It was seen a month later with coupling rods removed in preparation for its last journey to J. McWilliam of Shettleston where it was scrapped in the summer of 1966.

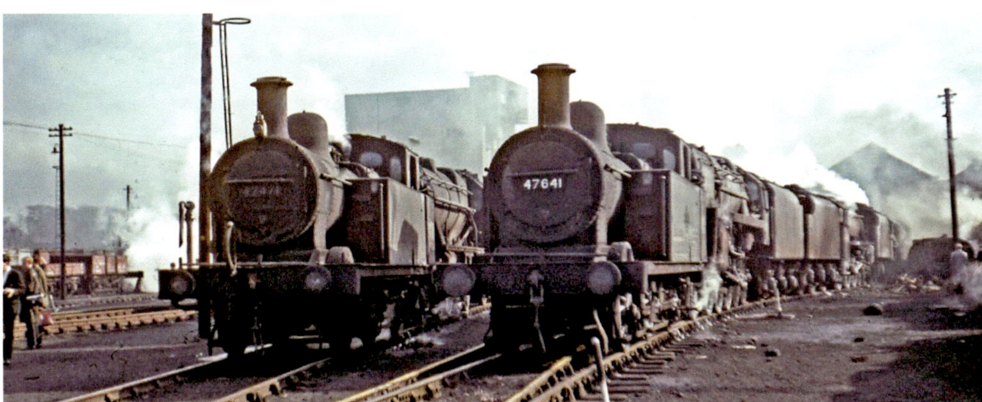

May 1966 LMS Jinty 3F 0-6-0Ts 47471 and 47641 Carlisle Kingmoor MPD.
LMS Jinty 3F 0-6-0Ts 47471 and 47641 head their respective lines of locomotives in steam and are preparing for their next duties. 47471 had been built by Vulcan Foundry for the LMS in January 1928 and had spent many years based at Lancaster Green Ayre depot, before it was transferred to Carlisle Kingmoor MPD in September 1960. Along with 47641, which has already been described, 47471 was withdrawn in December 1966. Both were scrapped by J. McWilliam of Shettleston in the spring of 1967.

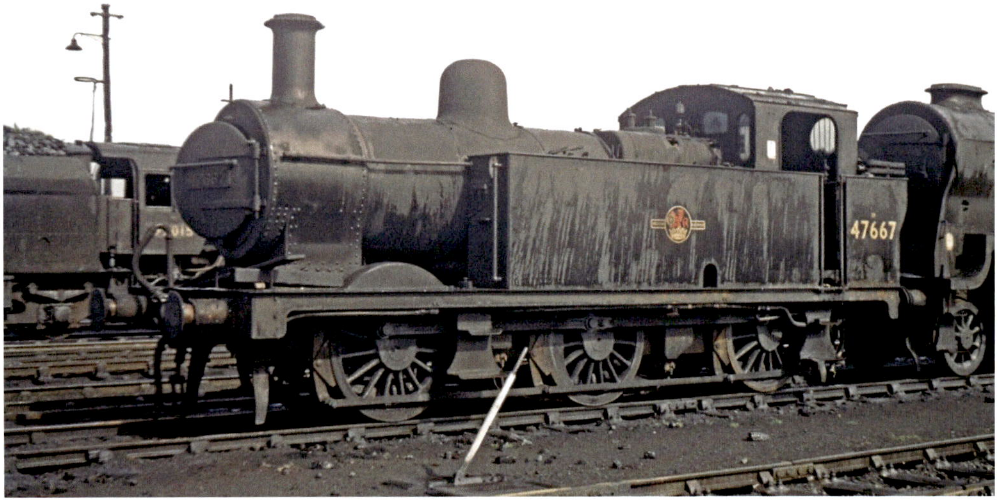

May 1966 LMS Jinty 3F 0-6-0T 47667 at Carlisle Kingmoor MPD.
LMS Jinty 3F 0-6-0T 47667 was built at Horwich Works for the LMS in April 1931. It spent much of its working life in the London area, particularly at Camden MPD. In April 1958 it was reallocated to Carlisle and spent time working from both Upperby and Kingmoor depots. It was reasonably clean when photographed on Kingmoor depot but was transferred away in September 1966 to Skipton and withdrawn in November. T. W. Ward at Killamarsh scrapped it in March 1967.

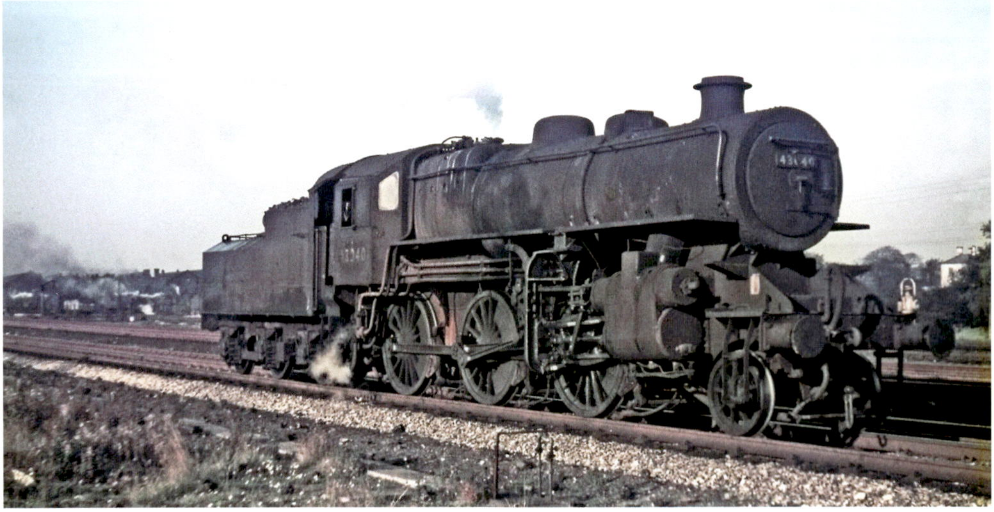

June 1966 LMS Ivatt 4MT 2-6-0 43040 at Carlisle Kingmoor MPD.
LMS Ivatt 4MT 2-6-0 43040 had been built for British Railways at Horwich Works in July 1949. It spent its first nine years of service allocated to Nottingham MPD. It was then transferred to Derby, Saltley and Trafford Park before it arrived at Carlisle Kingmoor in July 1964. It was to stay at Carlisle until one last move took it to Blyth North in August 1966. It was seen running past Kingmoor depot a few weeks before its move to the North East. It was withdrawn in November 1966 and scrapped by Clayton and Davie at Dunston-on-Tyne in the spring of 1967.

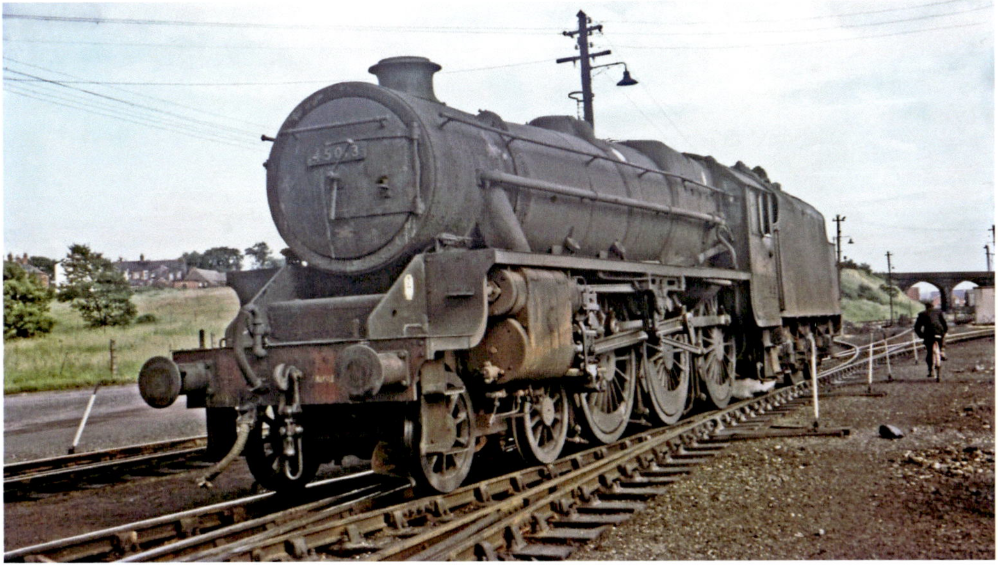

30 June 1966 LMS Black Five 4-6-0 45013 at Carlisle Kingmoor MPD.
LMS Black Five 4-6-0 45013 had been built at Crewe Works in April 1935 as one of the first batch of Black Fives to be constructed. It spent much of its existence allocated to Carlisle Kingmoor, although it did venture away to Crewe South for two years. It returned to Kingmoor in September 1951 and remained faithful to the depot until it was closed in December 1967. Unlike the majority of the remaining allocation, 45013 was not withdrawn but was reallocated in January 1968 to Stockport Edgeley depot. It was withdrawn in May 1968. It was still paired with a dome-less boiler, which was unique to the original batch of locomotives. In late 1968 it was cut up by T. W. Ward at Beighton a few months after steam had become a memory on British Railways.

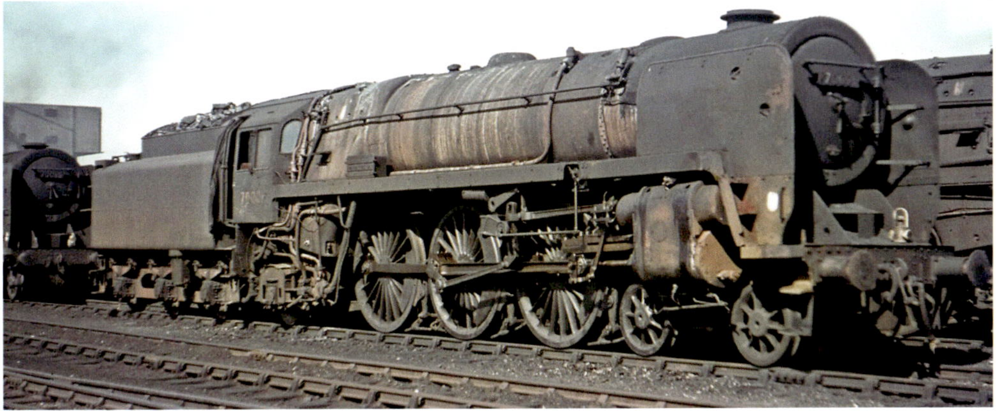

July 1966 BR Standard Britannia 7MT 4-6-2 70001 at Carlisle Kingmoor MPD.
BR Standard Britannia 7MT 4-6-2 70001 *Lord Hurcomb* had been built at Crewe in February 1951. It spent the first four years of service in East Anglia working from Stratford, Norwich and March before dieselisation in the area left it without any work. In March 1963 it moved to Willesden MPD and after a couple months moved on again to Aston depot. In October 1964 it found its final home, along with many other members of the class, when it was reallocated to Carlisle Kingmoor. In September 1966 it was withdrawn and before the end of the year it had been cut up by Motherwell Machinery and Scrap at Wishaw.

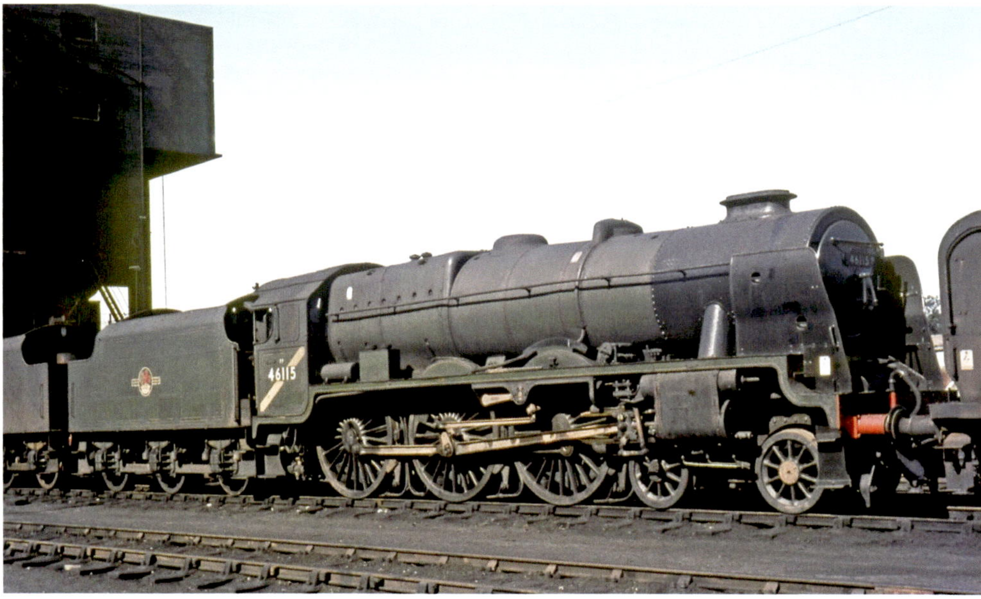

21 July 1966 LMS Royal Scot 4-6-0 46115 stored at Carlisle Kingmoor MPD.
LMS Royal Scot 4-6-0 46115 *Scots Guardsman* was withdrawn from service in January 1966 and purchased for preservation by Bob Bill. It was moved to the Keighley and Worth Valley Railway but was considered unsuitable for use on the line at that time. In May 1969 it was moved to the Dinting Railway Centre. After the death of Bob Bill, the locomotive was sold to a group and moved to Tyseley. It was finally sold to David Smith at Carnforth where it was restored to main-line condition. It has since worked special trains on the main-line network as well as visiting several heritage railways.

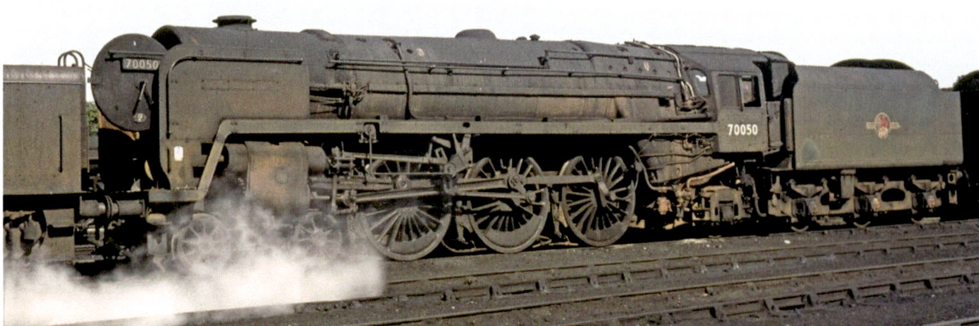

21 July 1966 BR Standard Britannia 7MT 4-6-2 70050 at Carlisle Kingmoor MPD.
BR Standard Britannia 7MT 4-6-2 70050 *Firth of Clyde* had been completed at Crewe Works in August 1954 and spent eight years allocated to Scottish Region depots. It was first at Polmadie and then on to Corkerhill before heading to Crewe North for three years and then three months at Banbury MPD. It was a late reallocation to Kingmoor depot in January 1966 and was withdrawn in August the same year. The photograph, taken a few weeks before its withdrawal, shows work being undertaken on its left-hand cylinder. This may have been the reason for its final demise after only twelve years of service.

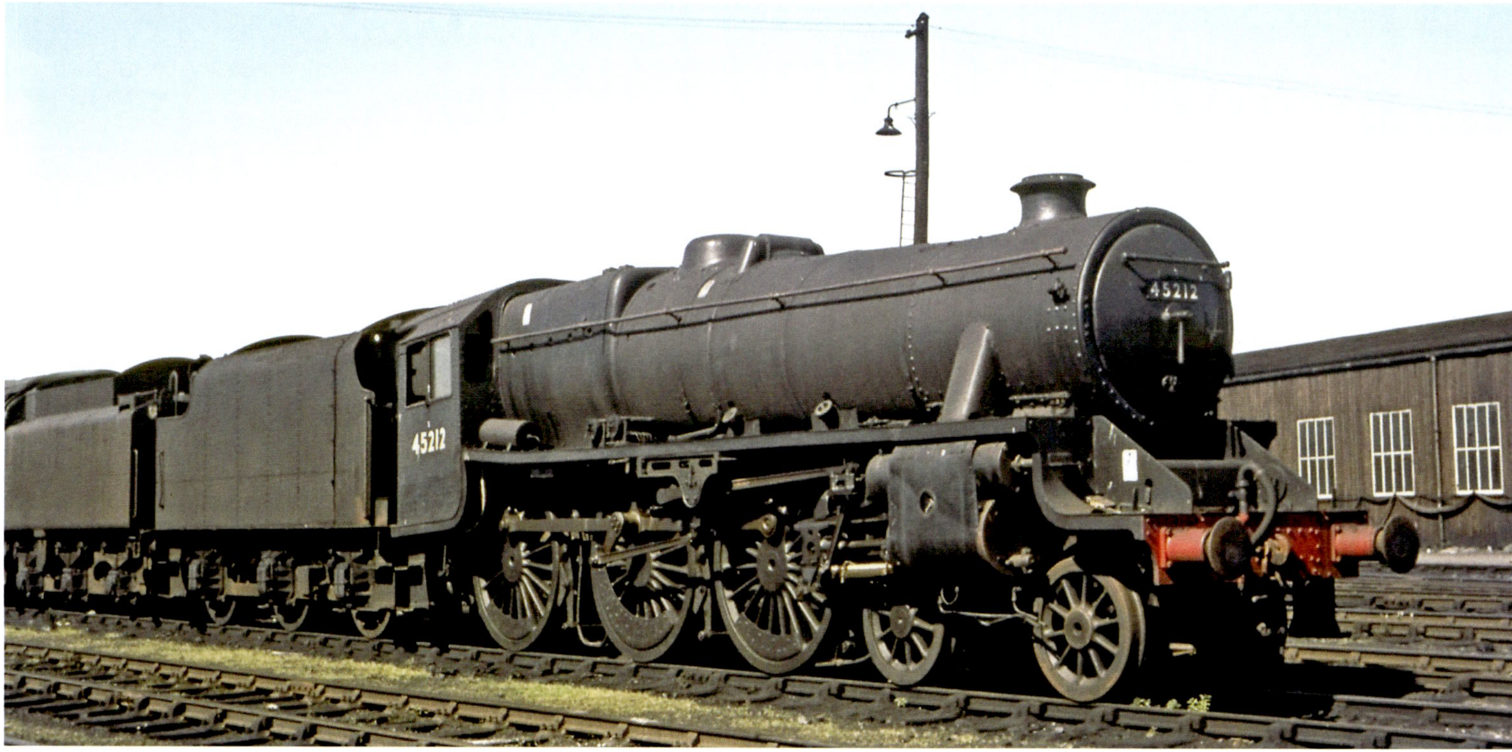

21 July 1966 LMS Black Five 4-6-0 45212 at Carlisle Kingmoor MPD.
LMS Black Five 4-6-0 45212 had been built for the LMS by Armstrong Whitworth of Newcastle and delivered in November 1935. It had spent most of its working life based at depots around the North West. At the formation of British Railways in January 1948 it was based at Fleetwood MPD and it spent sixteen years there before moving to Carnforth and on to Speke Junction depots. In September 1965 it was reallocated to Carlisle Kingmoor and in early 1966 was called to Crewe for a general overhaul. 45212 was photographed at Kingmoor after it had been recovered from a turntable pit and prepared for a return to Crewe for repairs. After the repairs the cab was left with a noticeable droop, but the locomotive ended its days at Lostock Hall MPD. It was withdrawn in August 1968 with the ending of steam on the national network. It was subsequently purchased privately for preservation on the Keighley and Worth Valley Railway. Some fifty years later it remains active on heritage lines and is regularly operated on the main line.

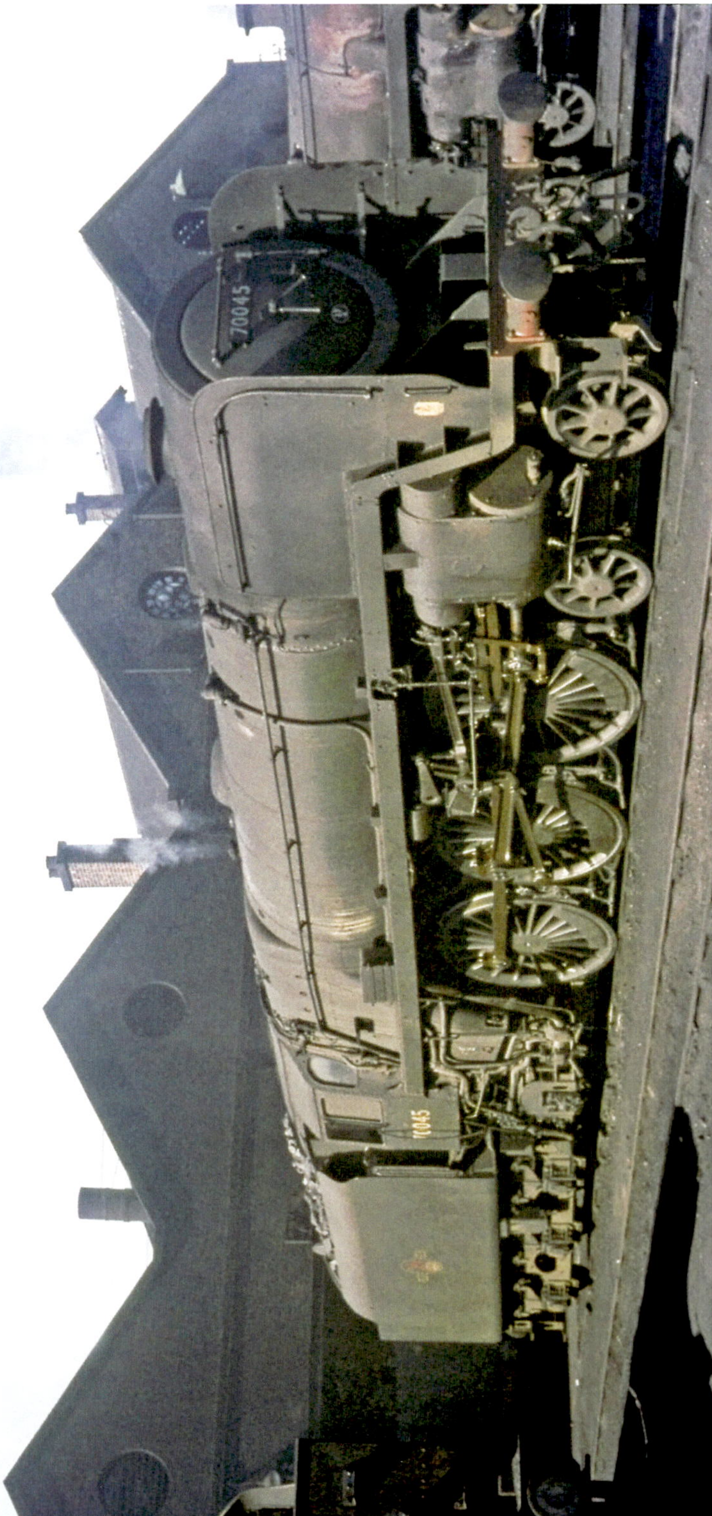

21 July 1966 BR Standard Britannia 7MT 4-6-2 70045 at Carlisle Kingmoor MPD.
BR Standard Britannia 7MT 4-6-2 70045 *Lord Rowallan* had emerged from Crewe Works in June 1954 and subsequently spent five years working from Holyhead. It then led a nomadic existence spending short periods at many depots from Crewe to Camden, from Newton Heath to Neasden, before returning to Holyhead for a further two and a half years of service. The magnetic draw of Britannias to Kingmoor took it there in January 1966 and it stayed until the depot closed in December 1967 when it was also withdrawn. The locomotive was then stored at Springs Branch, Wigan, for a couple of months before it moved to T. W. Ward of Beighton in March 1968 for cutting up.

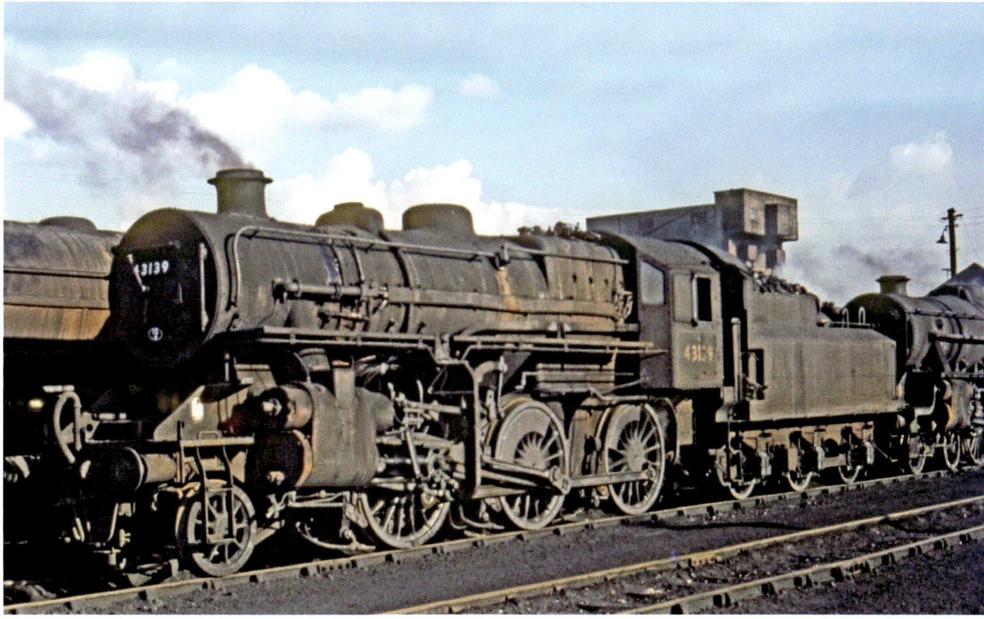

August 1966 LMS Ivatt 4MT 2-6-0 43139 at Carlisle Kingmoor MPD.
LMS Ivatt 4MT 2-6-0 43139 had been produced for British Railways by Doncaster Plant in July 1951; it was allocated new to Carlisle Canal MPD. With the closure of that depot, it moved across the city to Carlisle Kingmoor in June 1963. It was eventually withdrawn from there in September 1967 and was cut up by Motherwell Machinery and Scrap of Wishaw in February 1968.

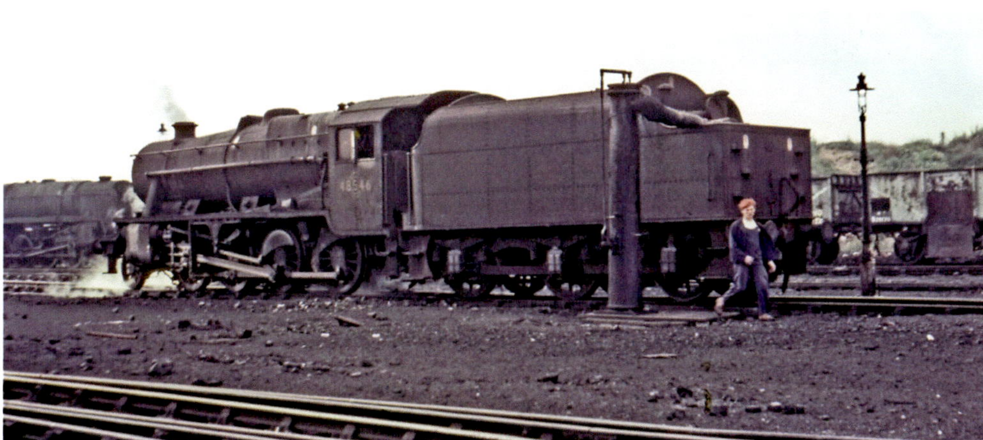

27 August 1966 LMS 8F 2-8-0 48546 at Carlisle Kingmoor MPD.
LMS 8F 2-8-0 48546 had been ordered by the Railway Executive Committee and built at Darlington Works. When completed in February 1945 it was loaned to the LNER and allocated to St Margaret's MPD. Under the ownership of British Railways, it spent nine years at Staveley Barrow Hill depot before moving to Patricroft. In March 1963 it was reallocated to Heaton Mersey and worked from there until the depot's closure took it to Lostock Hall to see out its last months of service. It was withdrawn in July 1968 days before the final elimination of steam on the national network.

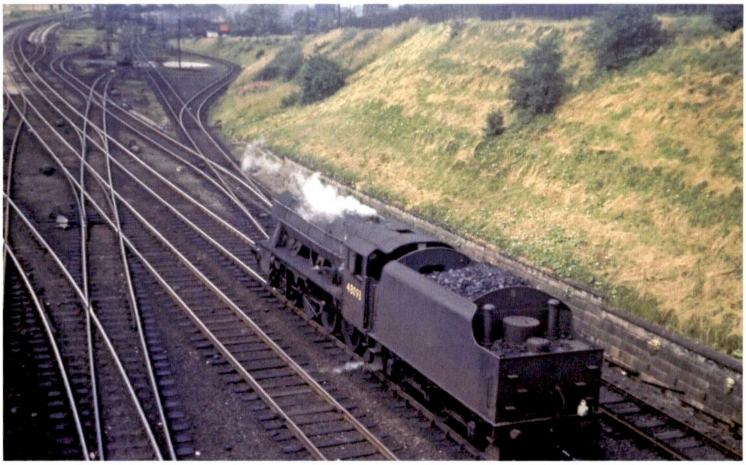

September 1966 LMS 8F 2-8-0 48093 at Carlisle Kingmoor.

LMS 8F 2-8-0 48093 had been built by the Vulcan Foundry for the LMS in February 1937. In 1941 it was one of fifty 8Fs requisitioned by the War Department for use in Persia. It was given the number WD621; however, it did not see service overseas because it was damaged in transit and was eventually repaired and returned to the LMS in 1943. Its entire British Railways career was spent in West Yorkshire, at Royston MPD until March 1965 when it was transferred to Stourton depot. With the closure of Stourton in January 1967 it returned to Royston until November of that year when it was withdrawn. It was seen negotiating the crossovers at Etterby Junction with the huge Kingmoor depot in the background.

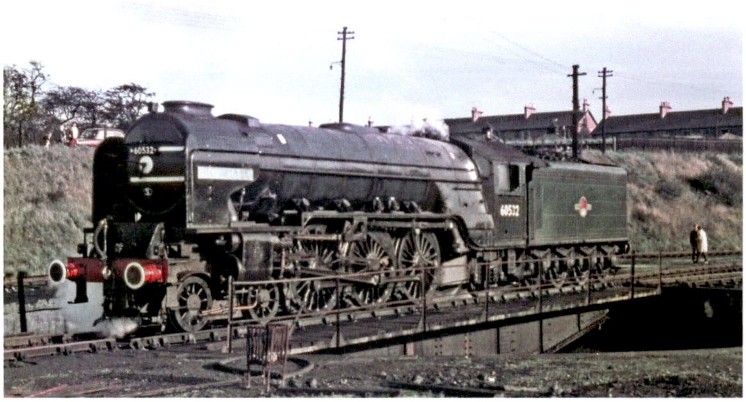

8 October 1966 LNER A2 4-6-2 60532 is turned at Carlisle Kingmoor MPD.

LNER A2 4-6-2 60532 *Blue Peter* had worked to Carlisle with an enthusiasts' special over the Waverley route from Edinburgh and was to return there via Beattock. The locomotive had been built at Doncaster Plant for the LNER but was actually delivered to British Railways in March 1948. After eighteen months' service from York MPD, it was reallocated to the Scottish Region with stints at Haymarket and Dundee Tay Bridge. It was allocated to Aberdeen Ferryhill for ten years of its active life and in 1966 became one of the last A2s in service. After withdrawal in December 1966, it was purchased privately for preservation and restored to working order in 1971. It did very little running and after years of inactivity it was returned to main line and heritage railway service in 1991.

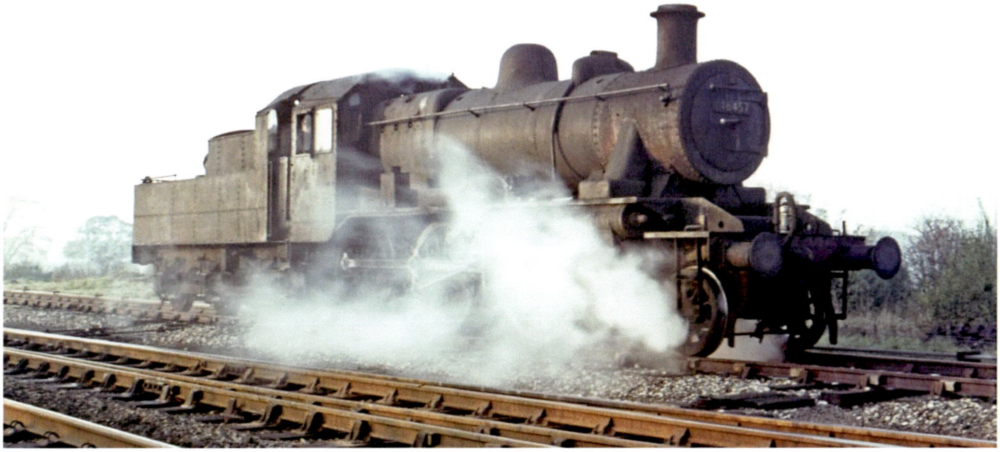

March 1967 LMS Ivatt 2MT 2-6-0 46457 near Carlisle Kingmoor MPD.
LMS Ivatt 2MT 2-6-0 46457 was built at Crewe Works and entered service with British Railways in May 1950. It was allocated to Workington MPD and remained there for eight years before spending four years at Carlisle Upperby. After stints at a number of depots including Bescot, Chester, Leamington and Tyseley it returned to Upperby in October 1966. With the closure of Upperby depot in December 1966 the locomotive transferred across to Kingmoor depot where it worked until its withdrawal in May 1967. It was cut up by McWilliam of Shettleston in November 1967.

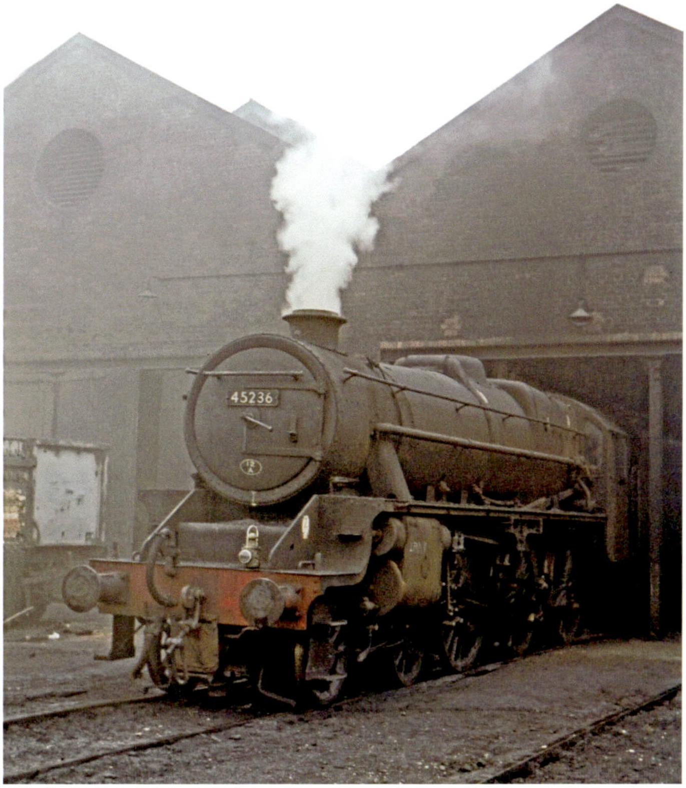

1 April 1967 LMS Black Five 4-6-0 45236 at Carlisle Kingmoor MPD.
LMS Black Five 4-6-0 45236 was built for the LMS by Armstrong Whitworth of Newcastle in August 1936. All of its service for British Railways was from depots in the North West. It had spent more than ten years at Barrow-in-Furness MPD before its reallocation to Carlisle Upperby in November 1959. In June 1963 it was moved across the city to Kingmoor depot where it continued to work until the depot was closed to steam in December 1967. 45236 was then withdrawn and in April 1968 was cut up by T. W. Ward at Inverkeithing.

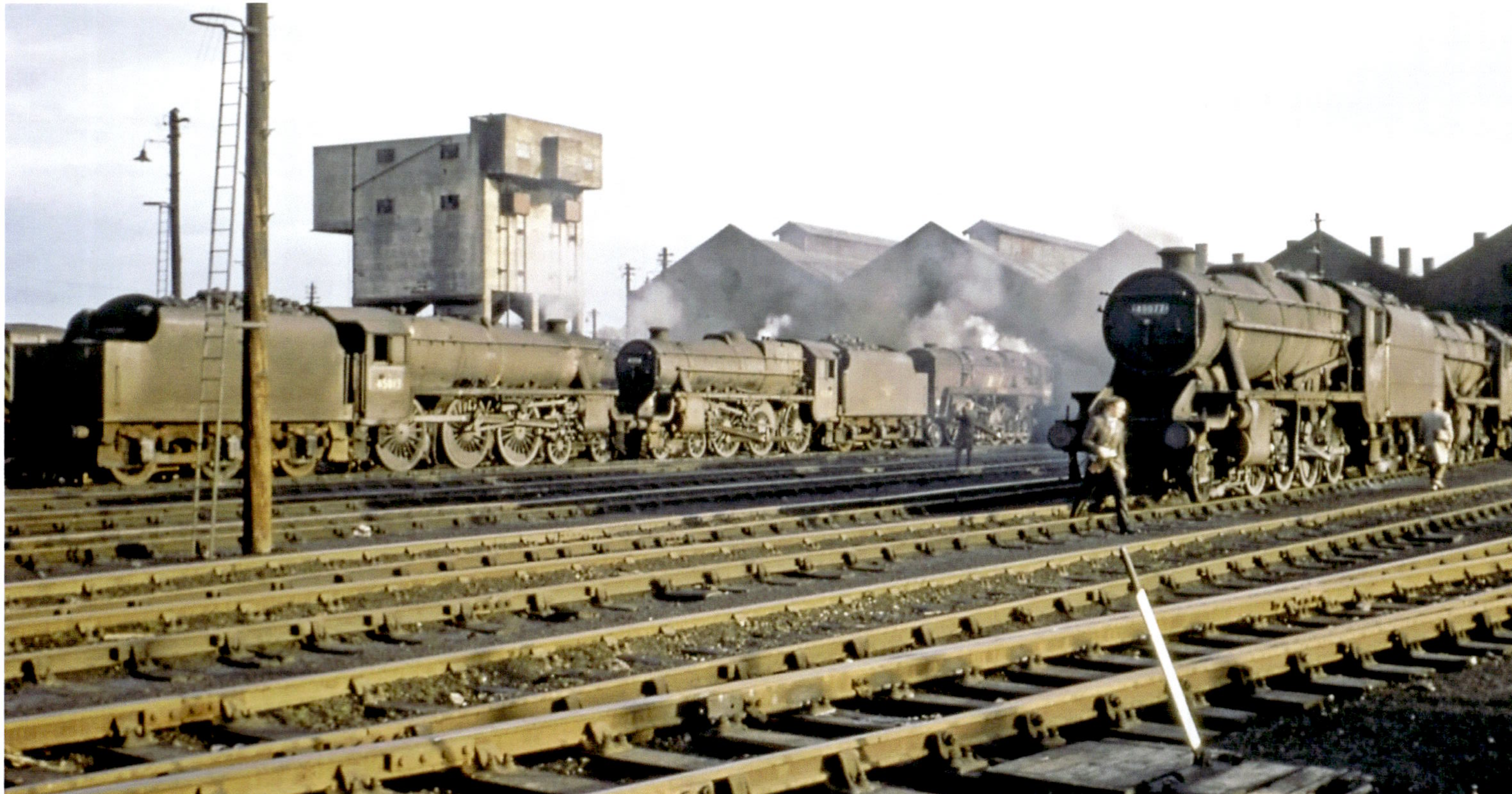

May 1967 LMS Black Five 4-6-0 45013 and LMS 8F 2-8-0 48077 on Carlisle Kingmoor MPD.
LMS Black Five 4-6-0 45013 and LMS 8F 2-8-0 48077 headed lines of locomotives being prepared for their next duties at Carlisle Kingmoor depot. At the beginning of 1967 there were 129 steam locomotives on the books at Kingmoor, including 45013 which had returned to the roster in September 1952 and remained faithful to the depot until its closure in December 1967. The engine was then transferred to Stockport Edgeley. With a little over six months of operation left, the depot was still servicing visiting locomotives, such as 48077 which had worked in from its home of Lostock Hall MPD.

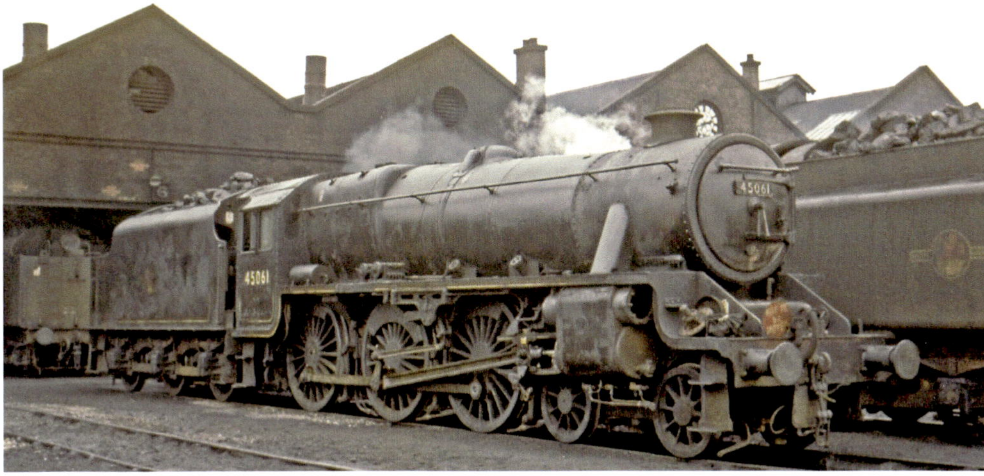

1 April 1967 LMS Black Five 4-6-0 45061 at Carlisle Kingmoor MPD.
LMS Black Five 4-6-0 45061 was built by Vulcan Foundry for the LMS in December 1934; it spent most of its working life at depots in the North West. It spent fifteen years allocated to Southport MPD before moving on to Aintree depot and eventually to Carlisle Kingmoor in 1966. It was seen on its home shed undergoing some work on its valves. It was fully coaled with huge lumps of coal and there were obvious signs that some cleaning of the locomotive had been attempted. It was withdrawn from Kingmoor in November 1967 and cut up by Motherwell Machinery and Scrap of Wishaw in March 1968.

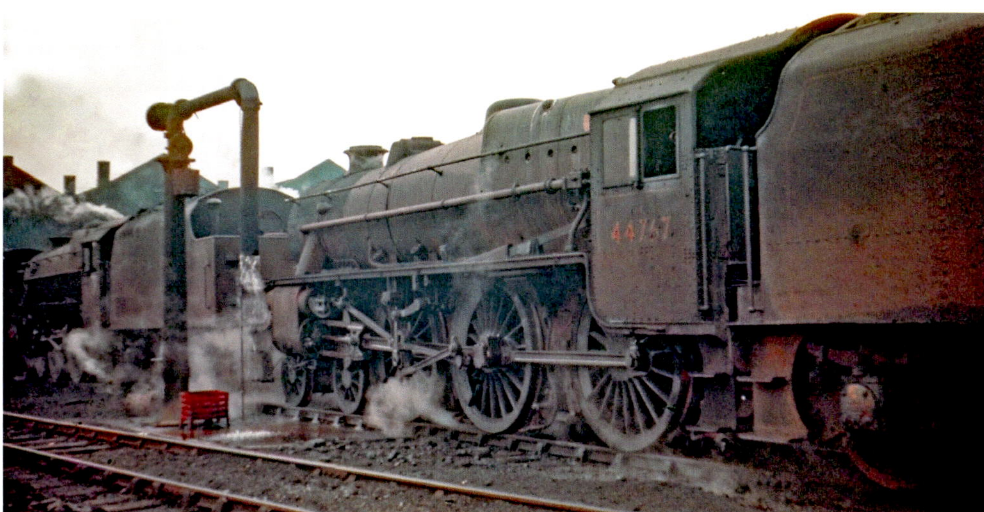

June 1967 LMS Black Five 4-6-0 44767 at Carlisle Kingmoor MPD.
LMS Black Five 4-6-0 44767 was built at Crewe Works at the very end of the existence of the LMS, and at the start of the nationalised British Railways. It was unique among the class of 842 locomotives as it was fitted with Stephenson link motion. Unusually as a 'one off' it remained in service for a considerable time – perhaps its reliability and its reputation for being a strong engine on the banks ensured it continued use, at a time when steam locomotives were withdrawn for relatively minor problems. With the closure of Kingmoor in December 1967 it was withdrawn and bought privately for preservation.

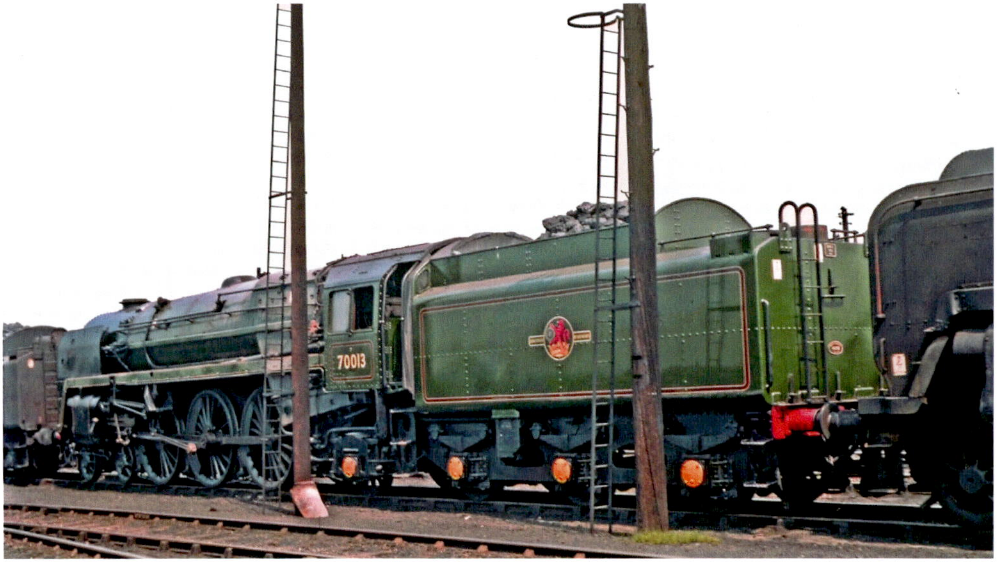

3 July 1967 BR Standard Britannia 7MT 4-6-2 70013 at Carlisle Kingmoor MPD.
BR Standard Britannia 7MT 4-6-2 70013 *Oliver Cromwell* had been outshopped from Crewe Works after a general overhaul in a special ceremony to mark the end of steam activities there. Four months later the engine was still in superb external condition with its lined green livery; it was such a marked contrast to the other working steam locomotives around it. 70013 was later selected to work the very last BR steam-hauled passenger train on 11 August 1968 and set aside for inclusion in the National Collection. It has regularly worked on the main line and several heritage railways since its return to steam in 2004 at the Great Central Railway.

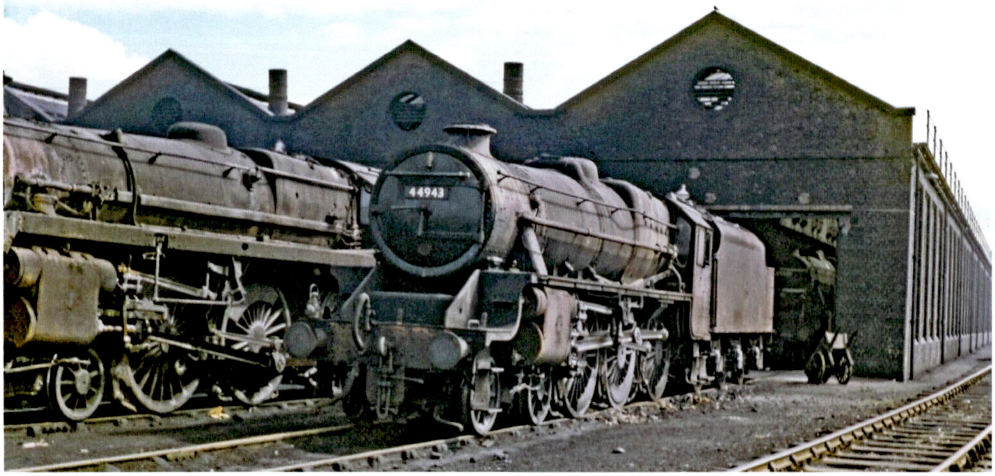

August 1967 LMS Black Five 4-6-0 44943 at Carlisle Kingmoor MPD.
LMS Black Five 4-6-0 44943 had been built at Horwich Works for the LMS in December 1946. It had spent its entire British Railways career based in West Yorkshire, mainly at Holbeck depot, with an allocation to Farnley MPD which lasted three years. It was a regular performer on the Settle and Carlisle line and was seen on Kingmoor after being prepared for its return home. It was eventually withdrawn a few weeks after this image in September 1967 and was cut up by T. W. Ward at Killamarsh in early 1968.

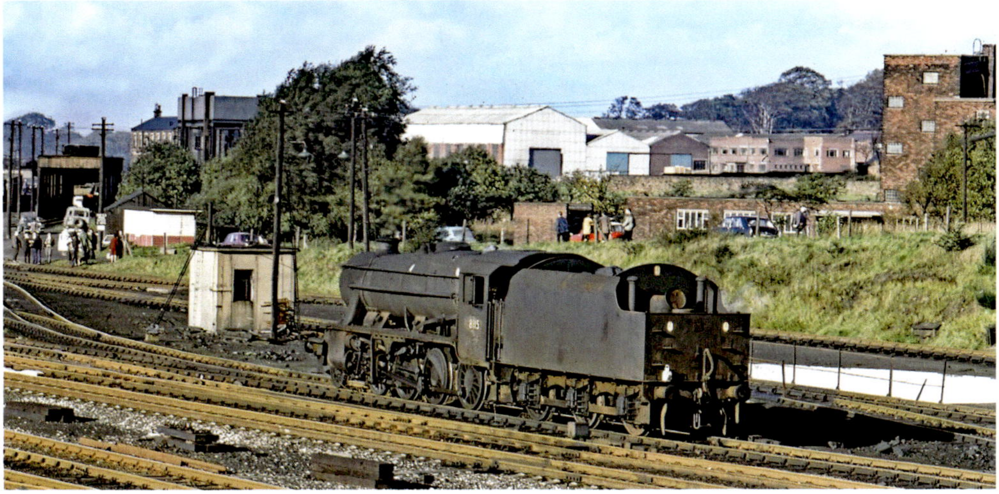

September 1967 LMS 8F 2-8-0 48115 at Carlisle Kingmoor MPD.

LMS 8F 2-8-0 48115 had been built at Crewe Works and entered service in April 1939 at Warwick MPD. With the dawn of British Railways, it had been reallocated to Westhouses depot and remained there for ten years before moving to Newton Heath in October 1958. It was to make the North West its home for the rest of its active life, at Patricroft MPD and then Heaton Mersey from March 1965. In May 1968 it was transferred to Rose Grove but two months later joined the lines of withdrawn locomotives there awaiting removal to scrapyards. 48115 was among the last batch of engines to leave Rose Grove in December 1968 and was cut up by Buttigieg's of Newport.

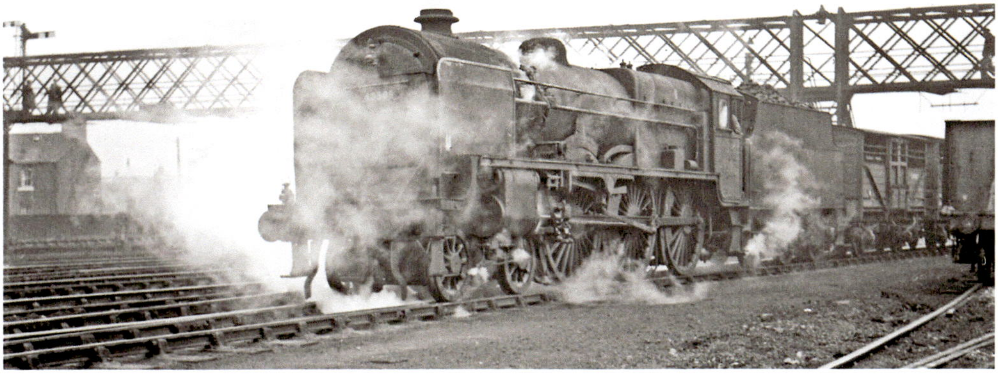

August 1961 LMS Patriot 4-6-0 45505 near Carlisle Upperby MPD.

LMS 4-6-0 Patriot 45505 *The Royal Army Ordnance Corps* slips as it gets its goods train underway near Carlisle Upperby shed. This loco had been paired with a modified high-sided Fowler tender and was in its last year of service. It had been built at Crewe Works in July 1932 and had spent a considerable amount of its career working from Preston and Carlisle Upperby depots. It had also spent ten years as a resident of Longsight but returned to Upperby in September 1960. In March 1962 it was one of four un-rebuilt Patriots reallocated to Lancaster Green Ayre but by May of that year it was in store at the closed remains of Preston MPD. It finally returned to Crewe Works in August 1962 for scrapping.

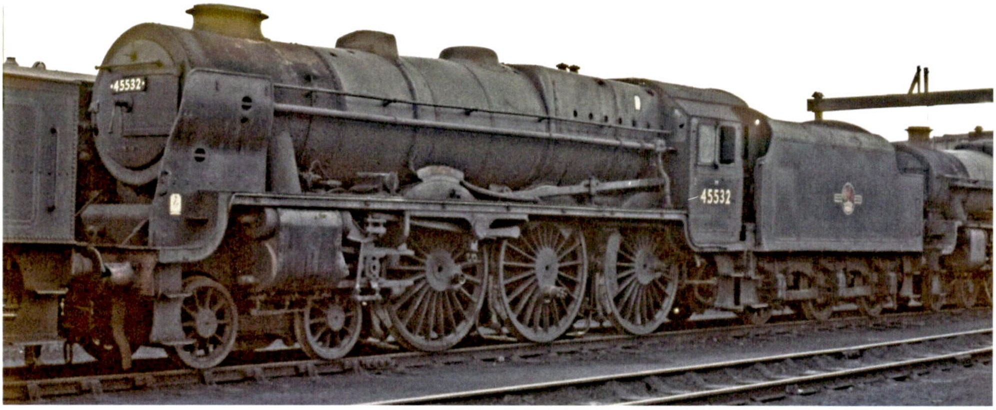

13 June 1964 LMS Patriot 4-6-0 45532 at Carlisle Upperby MPD.
LMS Patriot 4-6-0 45532 *Illustrious* had been built at Crewe in April 1933. It was rebuilt by British Railways in July 1948 with modifications designed by H. G. Ivatt. These involved the fitting of a large tapered boiler, new cylinders, new frame and a double chimney. The locomotive was allocated to Camden for nine years until November 1959 when it was reallocated to Nottingham MPD. It spent time at various depots including Saltley and Derby before it was moved to Carlisle Upperby. It was withdrawn in February 1964 and stored for a year before being cut up by Campbell's of Airdrie. (Photograph L. Flint)

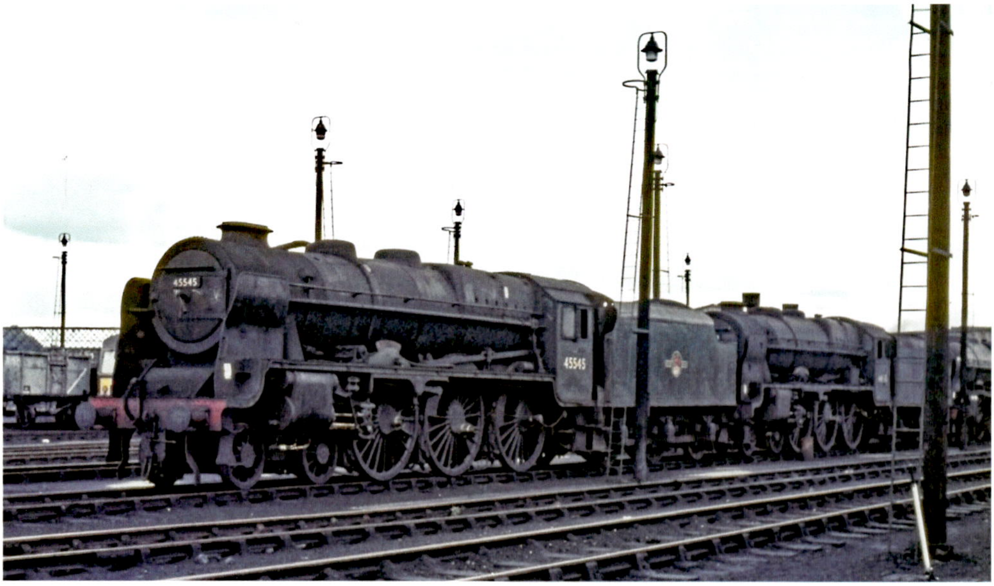

13 June 1964 LMS Patriot 4-6-0 45545 at Carlisle Upperby MPD.
LMS Patriot 4-6-0 45545 *Planet* had been built at Crewe in March 1934 to a Fowler design. In November 1948 it was rebuilt by British Railways after which it retained very little of the original locomotive. It had been Camden based throughout most of the 1950s until it moved to Crewe North in May 1957. Its final reallocation took it to Carlisle Upperby in June 1961. It was withdrawn from there in May 1964 and was cut up by Connell's of Coatbridge in December 1964. (Photograph L. Flint)

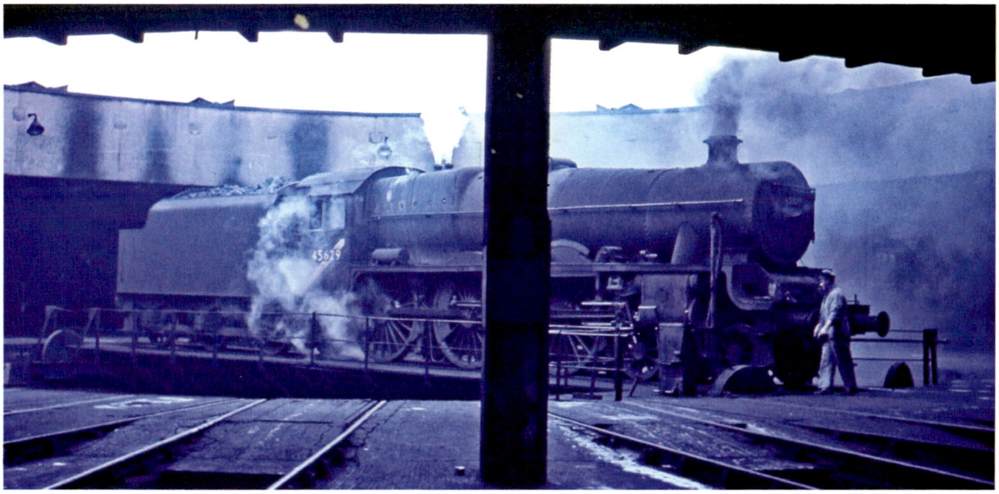

13 June 1964 LMS Stanier Jubilee 4-6-0 45629 in Carlisle Upperby MPD.
LMS Stanier Jubilee 4-6-0 45629 *Straits Settlement* was built at Crewe Works in November 1934 and had spent a considerable amount of time allocated to depots in the North West. It had been a Trafford Park locomotive before moving to Longsight in December 1956 and then Crewe North in June 1958. After a year at Carnforth 45629 was eventually transferred to Carlisle Kingmoor in June 1963. It was photographed on the turntable at Upperby shed while in its last year in service. Withdrawal came in May 1965 and in September of that year it was cut up by Motherwell Machinery and Scrap of Wishaw. (Photograph L. Flint)

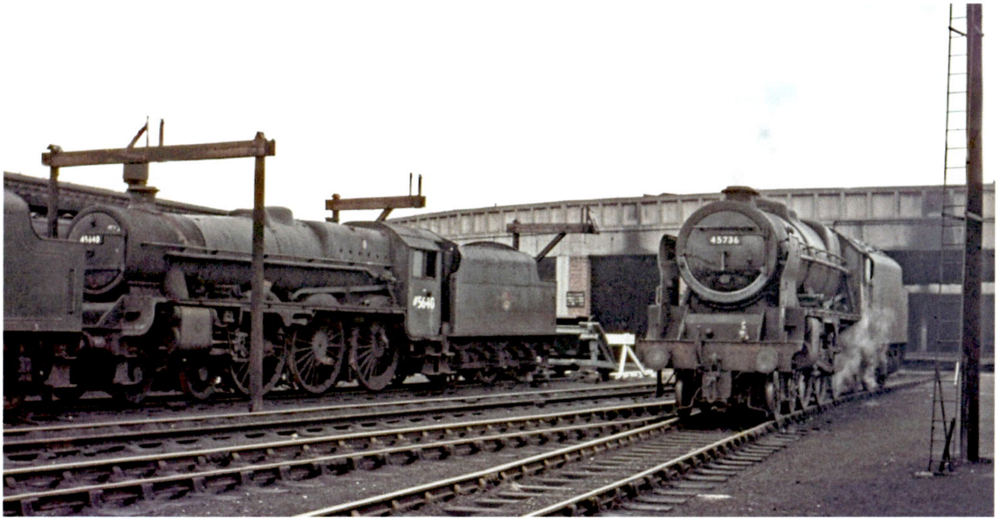

13 June 1964 LMS Stanier Jubilee 4-6-0s 45640 and 45736 at Carlisle Upperby MPD.
LMS Stanier Jubilee 4-6-0s 45640 *Frobisher* and 45736 *Phoenix* side by side at Upperby MPD. They illustrate the modifications made to 45736 when it was rebuilt in April 1942, with a larger boiler and a double chimney. The rebuild increased the power classification from 6P to 7P but only two Jubilees were actually modified, as it was decided to rebuild the Royal Scot and Patriot classes instead. 45640, which was reallocated to Carlisle Kingmoor in October 1952, had already been withdrawn in March 1964 and was cut up by Campbell's of Airdrie in early 1965. (Photograph L. Flint)

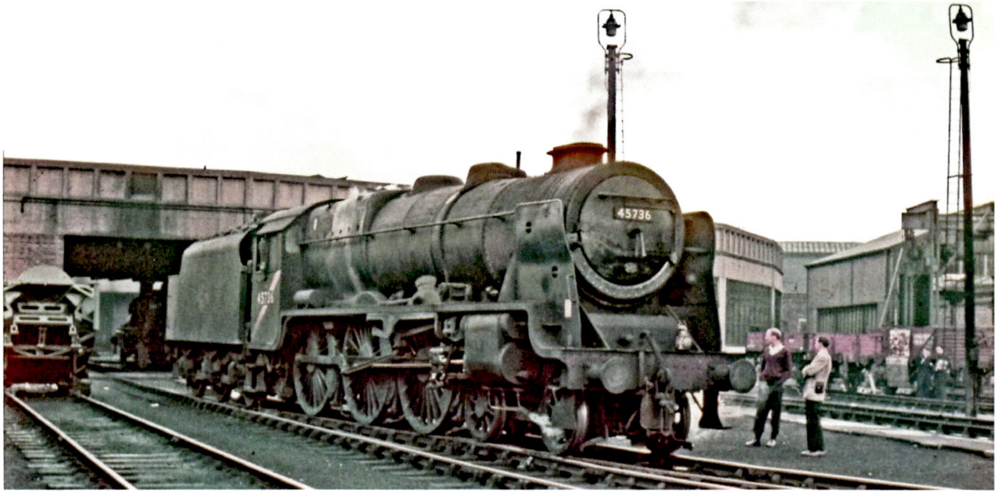

13 June 1964 LMS Stanier Jubilee 4-6-0 45736 at Carlisle Upperby MPD.
LMS Stanier Jubilee 4-6-0 45736 *Phoenix* had been built at Crewe in November 1936 but only five and half years later, along with 45735, it was rebuilt with a larger boiler and a double chimney. It spent six years at Camden until October 1954 when it was transferred to Crewe North, which was the start of its nomadic existence; Holyhead, Longsight, Camden, Willesden and Crewe were all its home and most more than once. In September 1963 it was reallocated to Carlisle Kingmoor but by December had moved again to Annesley. By June 1964 it had returned to Carlisle, this time to Upperby depot. It was withdrawn shortly afterwards in September and cut up by Hughes Bolchow's in North Blyth in March 1965. (Photograph L. Flint)

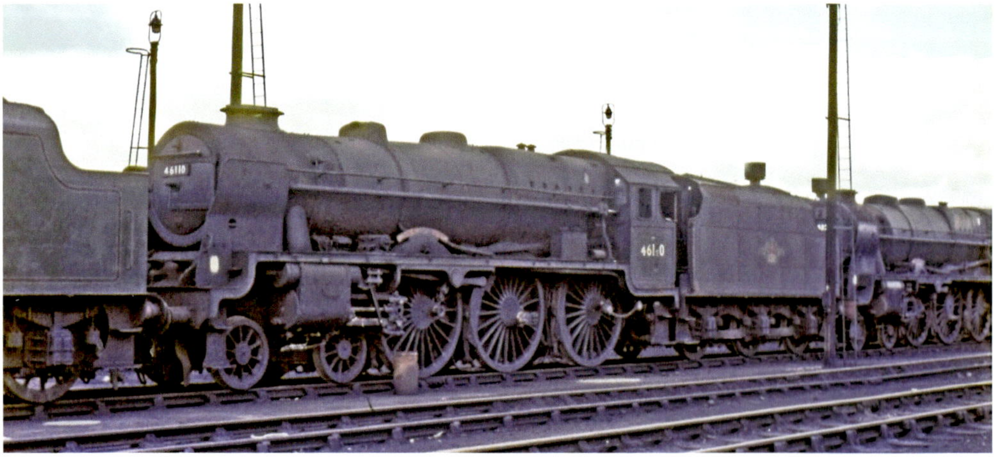

13 June 1964 LMS Royal Scot 4-6-0 46110 withdrawn at Carlisle Upperby MPD.
LMS Royal Scot 4-6-0 46110 *Grenadier Guardsman* had been built by the North British Locomotive Company of Glasgow for the LMS in September 1927. In January 1953 it was rebuilt with a large tapered boiler, a double chimney and new cylinders to a design by Stanier. Rebuilding of the class had commenced in 1943 with the last one being dealt with in 1955. 46110 had been based at Carlisle Upperby at the beginning of its British Railways career but then moved to numerous depots across the London Midland Region, before returning to Carlisle at Kingmoor in July 1963. Within six months of its return, it had been withdrawn and was scrapped by J. McWilliam of Shettleston in November 1964. (Photograph L. Flint)

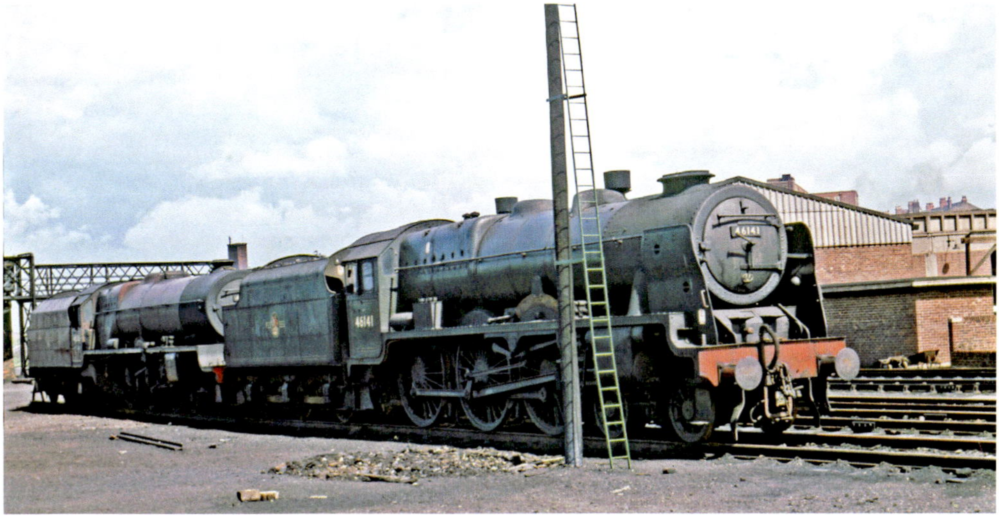

12 July 1964 LMS Royal Scot 4-6-0 46141 stored at Carlisle Upperby MPD.
LMS Royal Scot 4-6-0 46141 *North Staffordshire Regiment* had been built by the North British Locomotive Company of Glasgow for the LMS in November 1927. It was rebuilt at Crewe with a large tapered boiler in August 1950 and spent five years allocated to Carlisle Upperby until November 1959. In July 1962 it returned to Upperby depot until it was withdrawn in April 1964 and cut up at Crewe Works later in 1964.

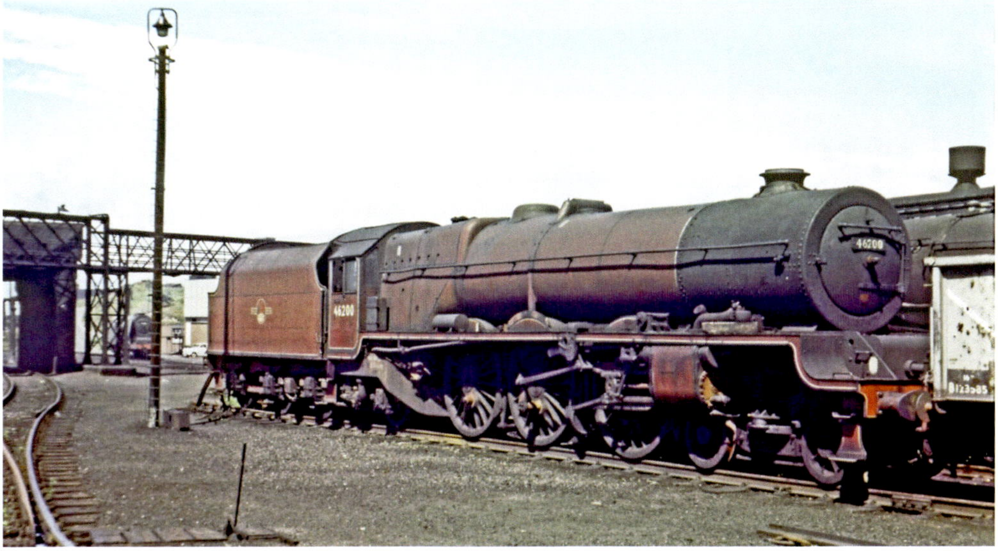

12 July 1964 LMS Princess Royal 4-6-2 46200 at Carlisle Upperby MPD.
LMS Princess Royal 4-6-2 46200 *The Princess Royal* had been built at Crewe in June 1933. It spent most of its working life allocated to Liverpool's Edge Hill depot. Over the years it strayed away from Edge Hill to both Camden and Crewe North on several occasions, but for only short periods of a few months. It left this familiar territory for good in September 1961 when it was transferred to Carnforth MPD and later Carlisle Upperby in January 1962, from where it was withdrawn in November 1962. A long period in store followed, with hints that it may be preserved, but it was finally scrapped by J. Connell's of Coatbridge in September 1964.

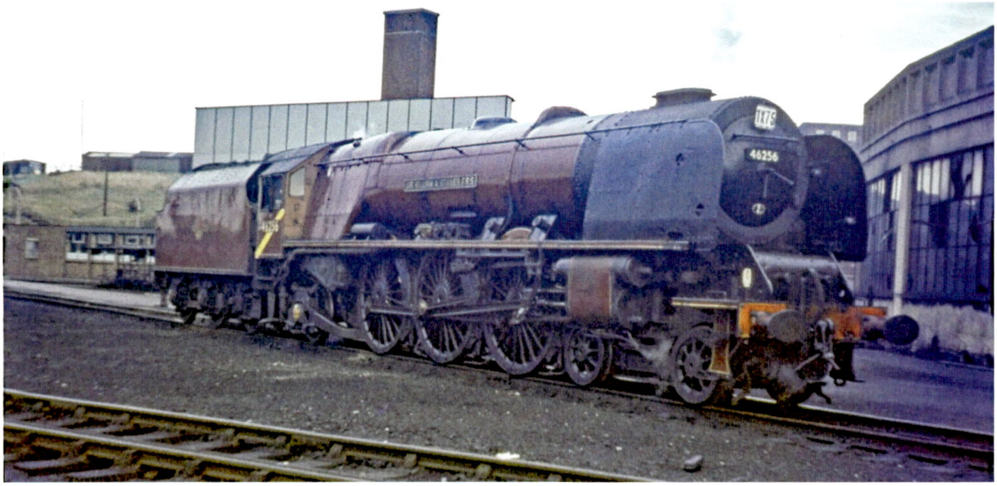

12 July 1964 LMS Coronation 4-6-2 46256 turned at Carlisle Upperby MPD.
LMS Coronation 4-6-2 46256 *Sir William A. Stanier FRS* had been retained in traffic for a month when the other last survivors of the class had been withdrawn 'en masse' at the beginning of September 1964. The reason for the reprieve was because it had been requested to work the RCTS special 'The Scottish Lowlander' between Crewe and Carlisle and the return journey. It was seen at Carlisle Upperby shed after being serviced and turned in preparation for the return round to Crewe, after which it was withdrawn.

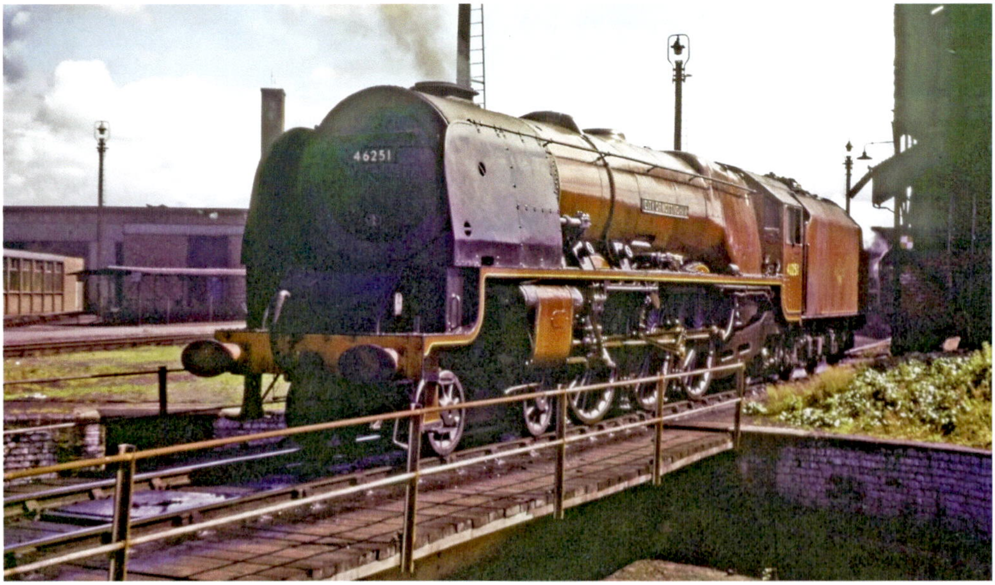

26 September 1964 LMS Coronation 4-6-2 46251 at Carlisle Upperby MPD.
LMS Coronation 4-6-2 46251 *City of Nottingham* had been built at Crewe Works in June 1944 and spent its working life on the West Coast Main Line. It had been based at Camden, Crewe North and Carlisle Upperby at different periods, with Camden being its home most often. Its last reallocation was to Crewe North in April 1963 and it was withdrawn from there in September 1964. It was seen being turned at Carlisle Upperby in the last few weeks of service. It had worked the Stephenson Locomotive Society's Pacific Pennine Rail Tour from Birmingham New Street to Carlisle.

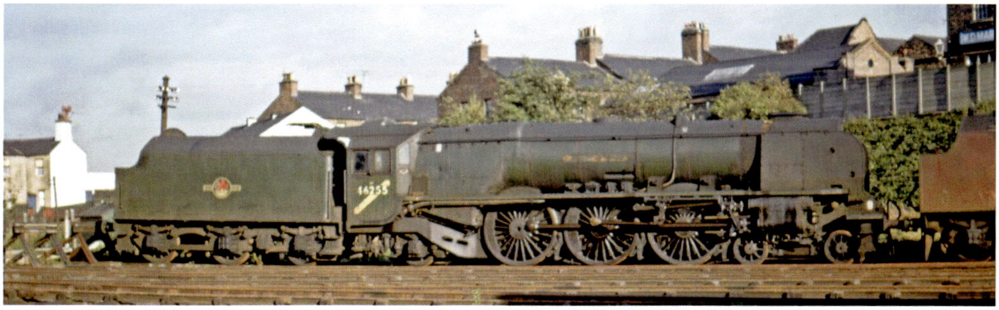

17 October 1964 LMS Coronation 4-6-2 46255 withdrawn at Carlisle Upperby.
LMS Coronation 4-6-2 46255 *City of Hereford* had been allocated to Carlisle Upperby in 1956 and on to Kingmoor in 1961, and so was very much a local engine often seen working in the city. It lasted in service until the last survivors were withdrawn together at the end of the summer timetable in 1964. It was seen after withdrawal, still in presentable condition but with nameplates removed, awaiting its final journey in early 1965 to the West of Scotland Shipbreaking of Troon.

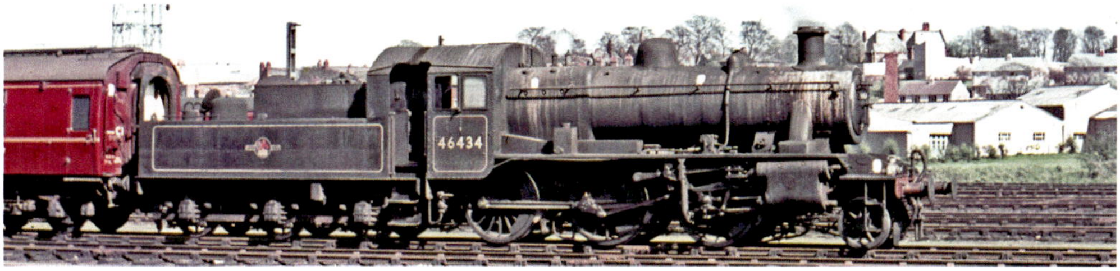

30 April 1966 LMS Ivatt 2MT 2-6-0 46434 near Carlisle Upperby MPD.
LMS Ivatt 2MT 2-6-0 46434 was built at Crewe Works for British Railways in December 1948. It was allocated to depots in the North West, spending time at Widnes and Springs Branch, Wigan. In May 1960 it was reallocated to Carlisle Upperby and remained there until withdrawal in September 1966. It was photographed in Upperby carriage sidings engaged on the tasks which had kept it busy for its final six years. It was cut up by Motherwell Machinery and Scrap of Wishaw in the summer of 1967.

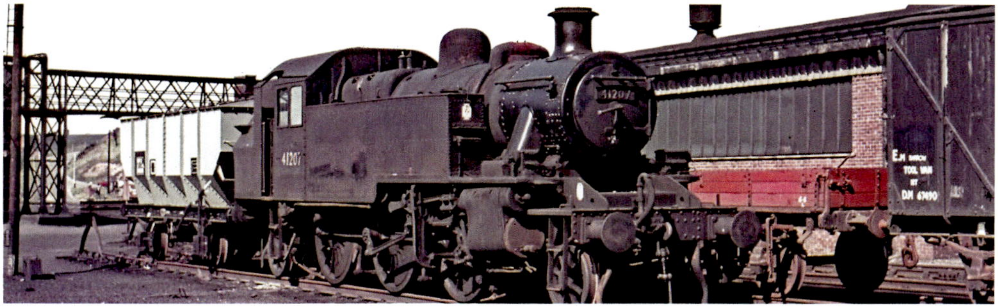

30 April 1966 LMS Ivatt 2MT 2-6-2T 41207 at Carlisle Upperby MPD.
LMS Ivatt 2MT 2-6-2T 41207 had been built at Crewe Works in December 1946 and had spent ten years allocated to Kentish Town and then Cricklewood depots. In June 1956 it was reallocated to Bristol Barrow Road MPD, which became its home for seven years, followed by two years at Shrewsbury. In the spring of 1966 it became a Carlisle Upperby engine but by the end of the year had been withdrawn. In March 1967 it had been sent to J. McWilliam of Shettleston and reduced to scrap metal.

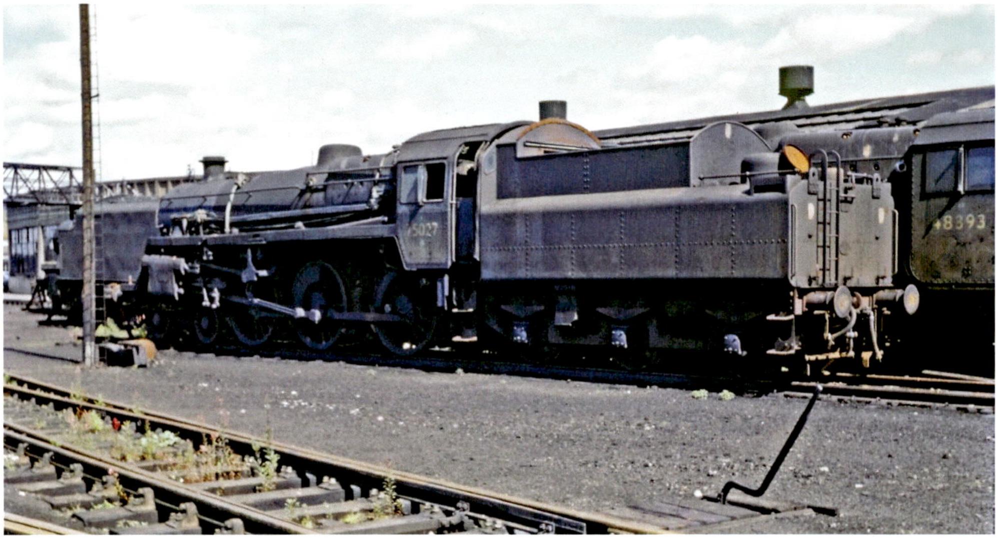

August 1967 BR Standard 4MT 4-6-0 75027 at Carlisle Upperby MPD.
BR Standard 4MT 4-6-0 75027 had been built at Swindon Works in May 1954. It was allocated to Western Region depots including Swindon, Reading and Oxford. In February 1960 it was sent to Templecombe and worked on the Somerset and Dorset line until November 1962 when it was transferred to Machynlleth. By August 1967 the prospects looked gloomy for 75027, seen here in a rusting and care-worn condition, together with many other stored locomotives at Carlisle Upperby. It was to take up banking duties from Tebay MPD and with the closure of that depot was reallocated to Carnforth. It continued to find work until the end of steam on British Railways in August 1968. It was purchased privately for preservation and donated to the Bluebell Railway where it arrived in early 1969. It has been a regular performer on that line since.

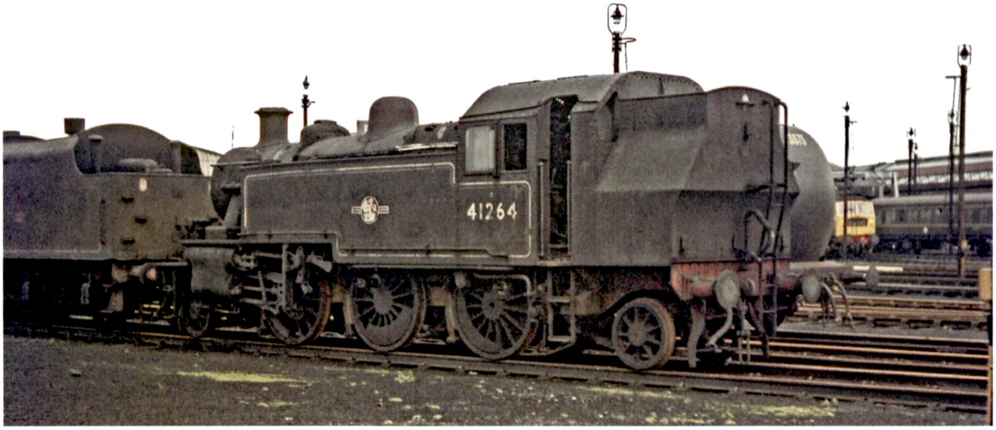

21 October 1967 LMS Ivatt 2MT 2-6-2T 41264 at Carlisle Upperby MPD.
LMS Ivatt 2MT 2-6-2T 41264 had been built at Crewe Works in July 1950. It spent its first four years of service allocated to Blackpool depot. In September 1954 it moved to West Yorkshire and became a resident of Low Moor MPD before reallocation to several other local depots. It spent a year and nine months in store at Low Moor before reinstatement in October 1963, with a transfer to Warrington and then on to Southport. It found its way to Carlisle Upperby in February 1965 and was withdrawn from there in December 1966.

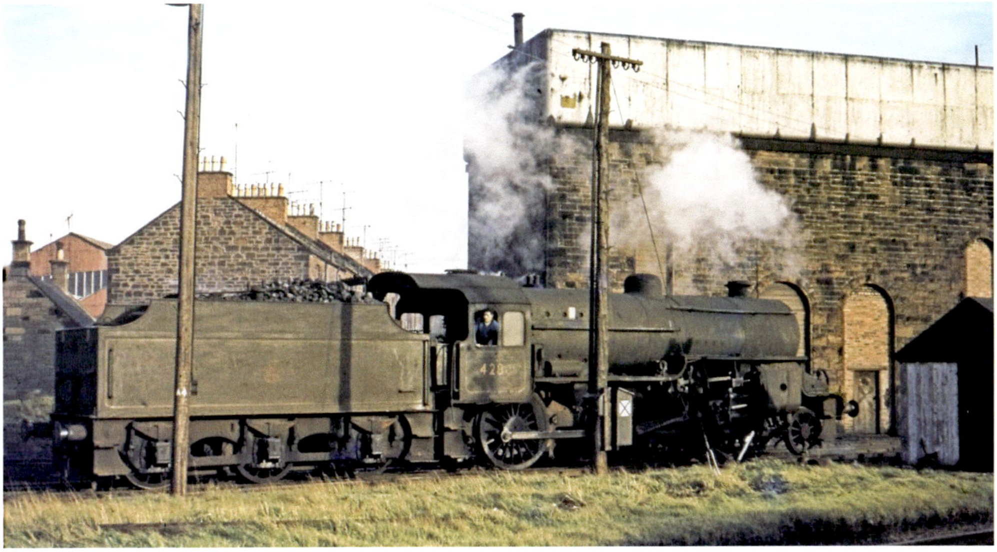

July 1962 LMS Crab 5P6F 2-6-0 42881 in Carlisle.
LMS Crab 5P6F 2-6-0 42881 was built at Crewe Works in June 1930. From the inception of British Railways in January 1948 it had spent its career at Carlisle Kingmoor. It was one of the early casualties of the class when it was withdrawn in November 1962, one of sixty-one Crabs to be scrapped that year. It was cut up at St Rollox Works at a time before scrapping locomotives more commonly became the domain of private yards.

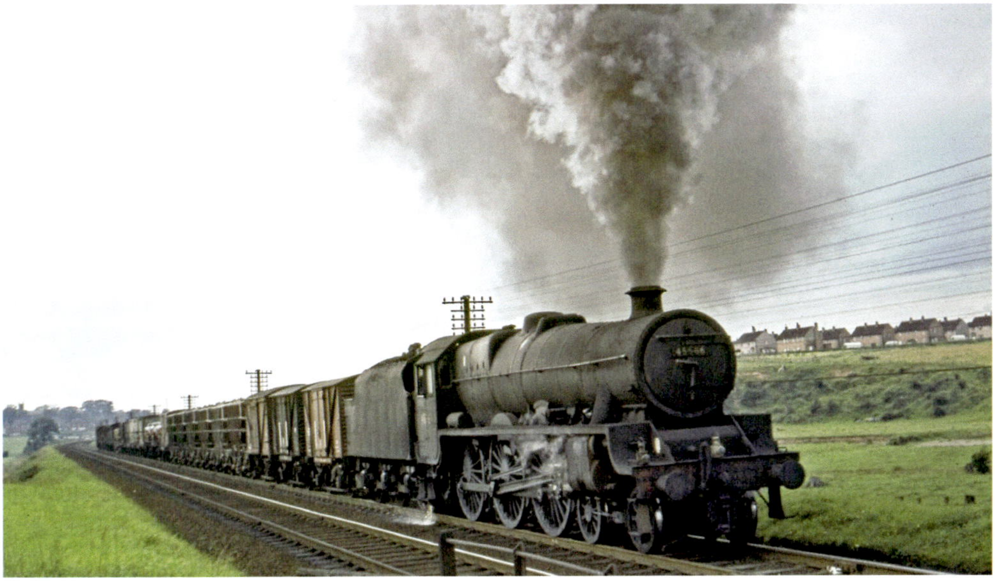

5 September 1963 LMS Jubilee 4-6-0 45586 near Carlisle.
LMS Jubilee 4-6-0 45586 *Mysore* had been built by the North British Locomotive Company of Glasgow for the LMS in December 1934. It was regularly allocated to Crewe North throughout its career. It spent various lengths of time at other depots including Edge Hill, Bushbury, Camden and Aston. In August it was to move one last time to Crewe South and was withdrawn from there in January 1965. It was seen struggling away from Carlisle with a heavy fitted goods train.

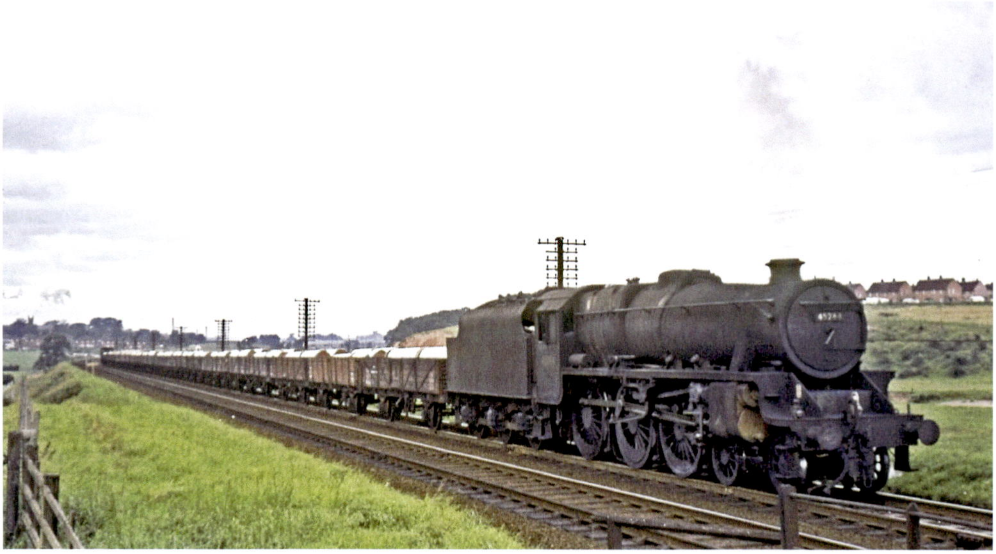

5 September 1963 LMS Black Five 4-6-0 45284 near Carlisle Upperby.
LMS Black Five 4-6-0 45284 had been part of the batch of the class built by Armstrong Whitworth of Newcastle for the LMS in 1936. It spent its early British Railways career based at depots in Manchester, latterly Newton Heath, before moving to Willesden in January 1960. It was reallocated to Edge Hill in October 1961 and it remained active from there until it was withdrawn in the final weeks of steam traction in 1968. It was seen leaving Carlisle with a long train of steel tubes. By the end of 1968 it had been cut up by T. W. Ward of Killamarsh.

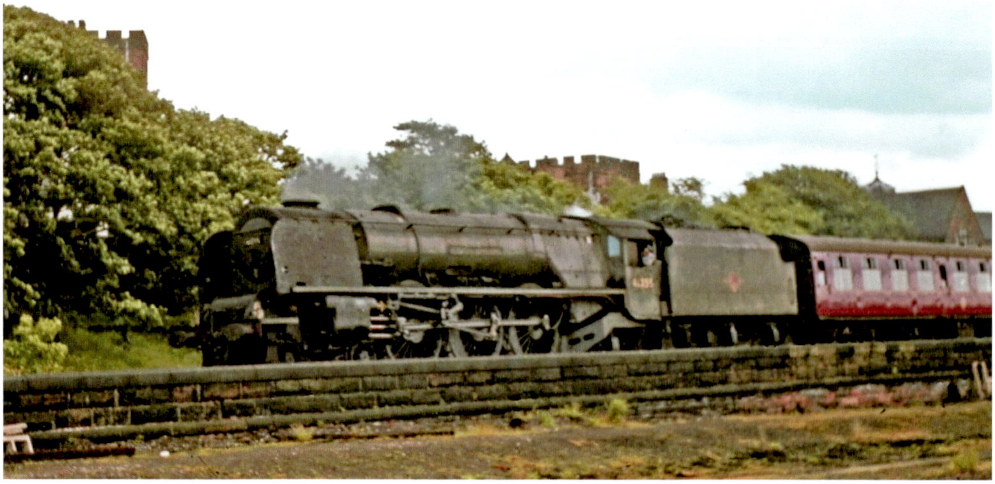

13 June 1964 LMS Coronation 4-6-2 46255 at Carlisle.
LMS Coronation 4-6-2 46255 *City of Hereford* was for many years a familiar sight around Carlisle. It had returned to the books at Upperby in February 1956 and in April 1961 was reallocated to Carlisle Kingmoor. It was among the survivors of the class which were all withdrawn on 12 September 1964 with the exception of 46256 which was retained in traffic for a couple of weeks in order to work a booked enthusiasts' special on 26 September.

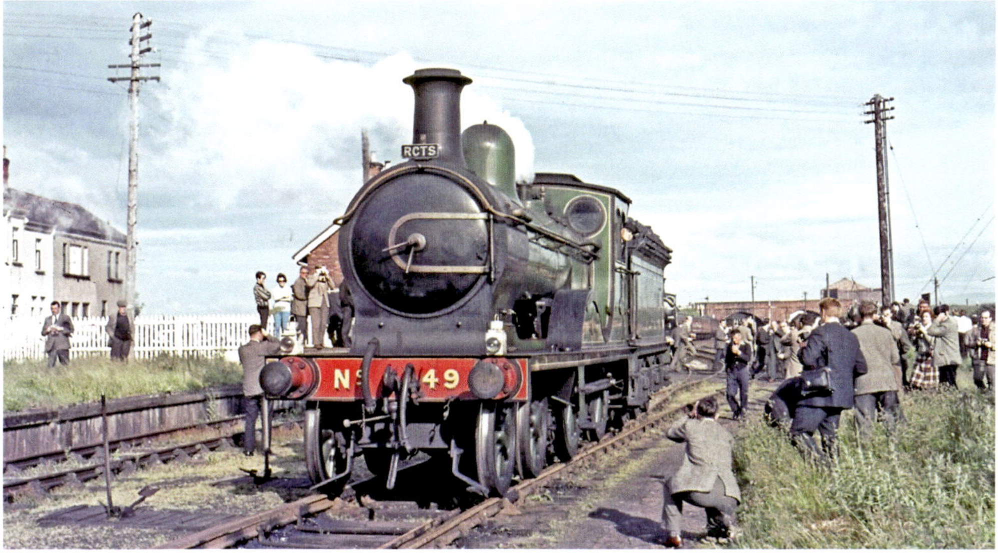

13 June 1964 GNSR 4-4-0 49 runs round at Silloth.
GNSR 4-4-0 49 *Gordon Highlander* had been built in 1920 by the North British Locomotive Company of Glasgow for the Great North of Scotland Railway; it was one of a class of eight mixed-traffic locomotives. It was withdrawn from ordinary service in June 1958 as BR D40 62277 and set aside for preservation. It remained active working special trains until 1966 before being placed in the Glasgow Museum of Transport. In 2011 49 was loaned to the Museum of Scottish Railways on the Bo'ness and Kinneil Railway. (Photograph L. Flint)

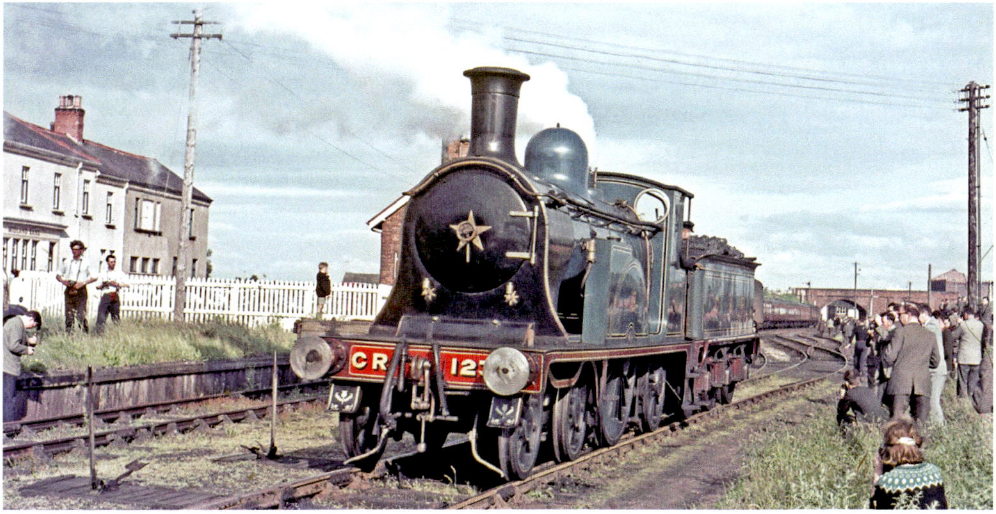

13 June 1964 CR 4-2-2 123 at Silloth.
123 was built by Neilson and Company, in collaboration with the Caledonian Railway, for the Edinburgh International Exhibition in 1886 where it was awarded a gold medal. It was the only single-driver locomotive operated by a Scottish railway and the last left in service in the British Isles. It was withdrawn in 1935 and preserved. In 1958 it was put back into working order to work enthusiasts' special trains. In 1965 it was retired from this work and placed in Glasgow Transport Museum. (Photograph L. Flint)

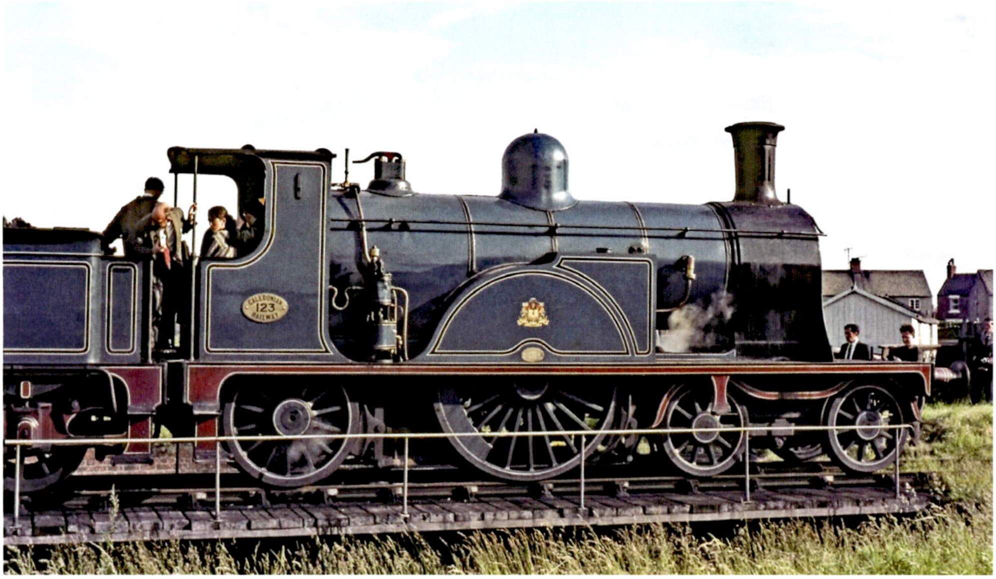

13 June 1964 CR 4-2-2 123 on the turntable at Silloth.
CR 4-2-2 123 was turned on the turntable at Silloth prior to running round the 'Solway Ranger' and heading the train back to Carlisle. The special had been organised by the West Riding branch of the RCTS and was hauled by SR Merchant Navy 4-6-2 35012 *United States Lines* from Leeds. (Photograph L. Flint)

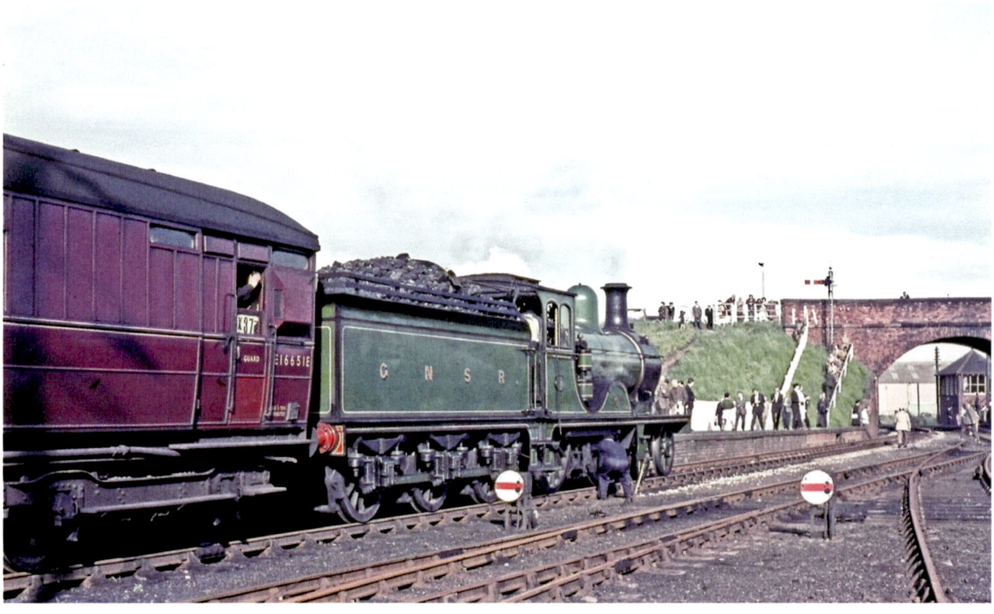

13 June 1964 GNSR 4-4-0 49 at Silloth.
GNSR 4-4-0 49 had been turned and run round its train the 'Solway Ranger' in readiness for the return journey to Carlisle Citadel. It awaited 123 to be coupled in front of it. The special had been organised by the RCTS and originated in Leeds and included a trip from Penrith to Workington hauled by LMS Ivatt 2MT 2-6-0s 46426 and 46458. (Photograph L. Flint)

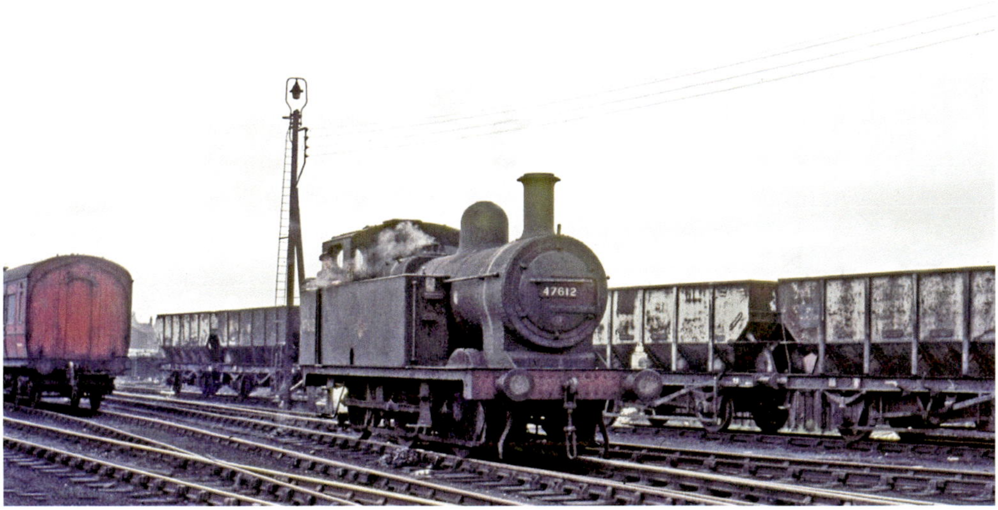

13 June 1964 LMS Jinty 3F 0-6-0T 47612 shunts at Workington.
January LMS Jinty 3F 0-6-0T 47612 was built by Beardmore of Dalmuir, Glasgow, in October 1929 for the LMS. It had been a long-term resident of Bescot MPD but at the formation of British Railways it had moved on to Northampton. Its next transfer was to Speke Junction in March 1959 before it finally arrived at Workington in November 1962. It was to remain active there until withdrawn in December 1966. It was scrapped by J. McWilliam of Shettleston in the summer of 1967. (Photograph L. Flint)

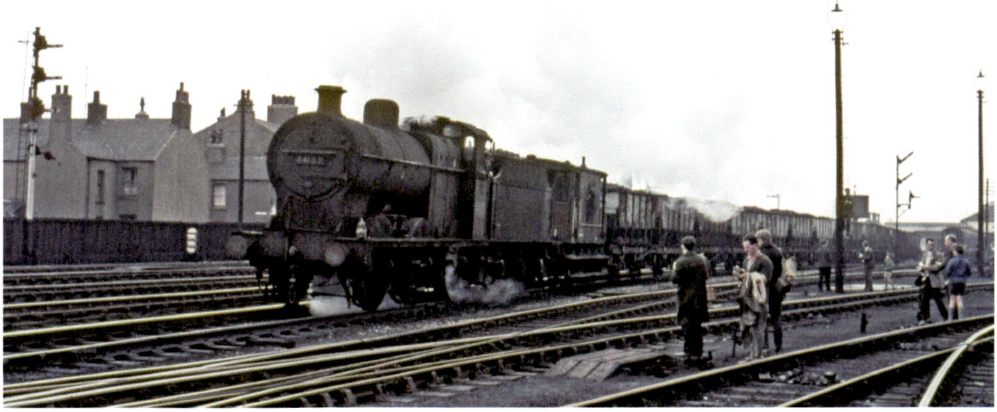

13 June 1964 LMS 4F 0-6-0 44157 trundles through Workington.
LMS 4F 0-6-0 44157 was built at Crewe Works in December 1925. At the dawn of British Railways in January 1948 it was allocated to Toton MPD. In December 1956 it moved on to Nuneaton and it became a Carlisle-based locomotive with its July 1958 move to Carlisle Canal depot. With the closure of Carlisle Canal in June 1963 it was transferred to Workington MPD where it continued with its toils until it was withdrawn in November 1965. It was cut up by Motherwell Machinery and Scrap of Wishaw. (Photograph L. Flint)

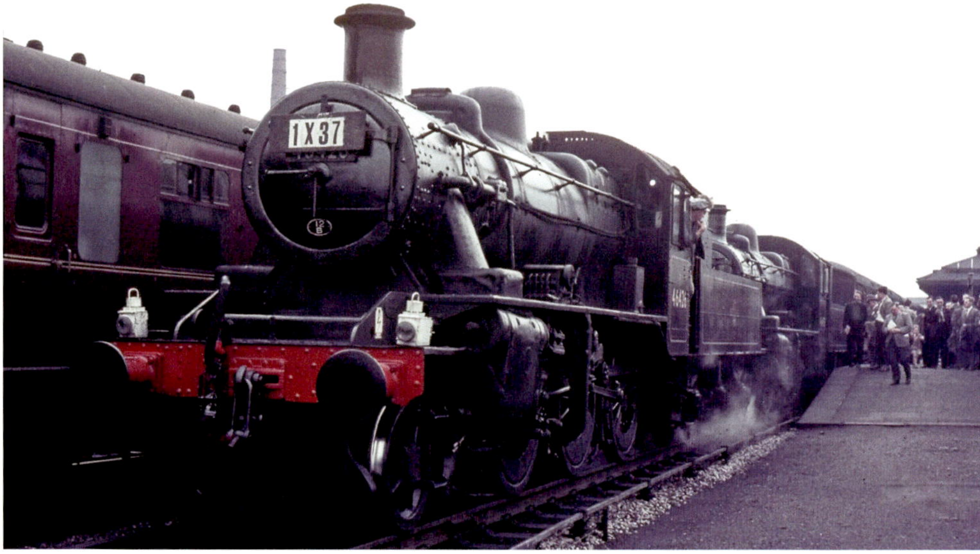

13 June 1964 LMS Ivatt 2MT 2-6-0s 46426 and 46458 on the 'Solway Ranger' at Workington.
LMS Ivatt 2MT 2-6-0s 46426 & 46458 arrived at Workington at the head of the Solway Ranger Rail Tour. The tour had been hauled from Leeds City to Penrith by Merchant Navy 4-6-2 35012. The two Ivatts had then headed the train from Penrith to Workington. The tour participants boarded a DMU set for a trip to Sellafield and then on to Carlisle while the Ivatts headed to Carlisle with the empty stock.

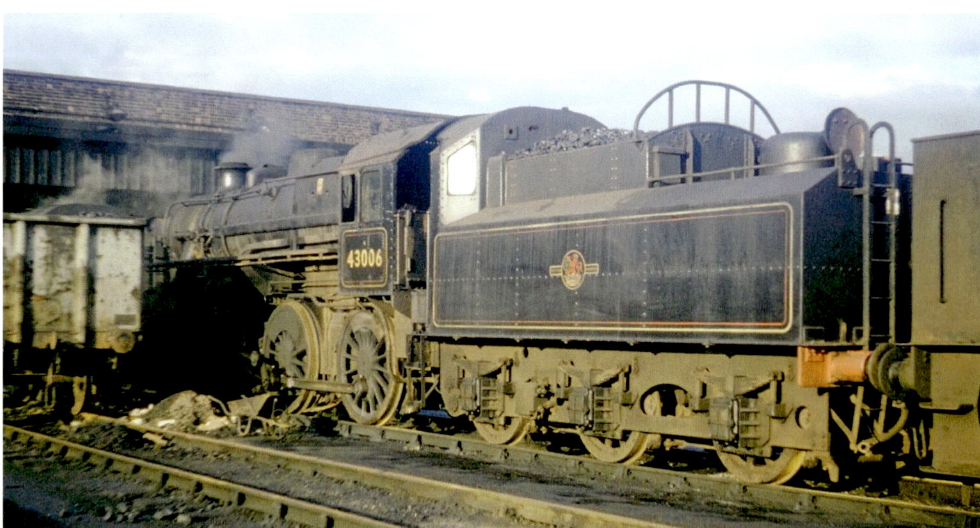

August 1966 LMS Ivatt 4MT 2-6-0 43006 at Workington MPD.
LMS Ivatt 4MT 2-6-0 43006 had been built at Horwich Works for British Railways and entered service in February 1948. It was allocated to Workington MPD from new and remained there until the depot closed to steam in January 1968. It then moved to Lostock Hall but was withdrawn in March 1968 and scrapped later that year by Arnott Young of Dinsdale.

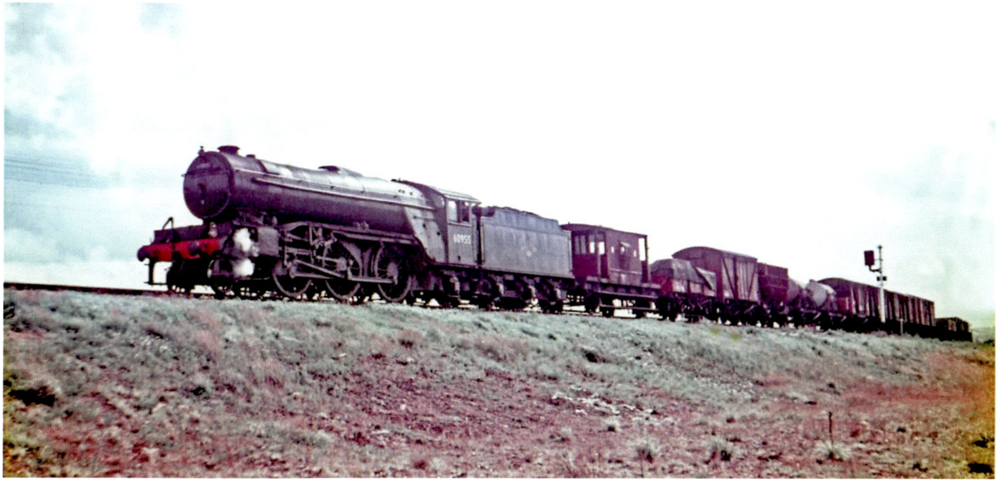

September 1964 LNER V2 2-6-2 60955 near Carlisle.
LNER V2 2-6-2 60955 had always been based in Scotland throughout its service and after transfer to St Margaret's depot in Edinburgh it became a frequent visitor to Carlisle. It regularly worked over the Waverley route with heavy mixed-goods trains, as here, near Carlisle. It continued to work the line until September 1966 when it was withdrawn and scrapped By Campbell's of Airdrie.

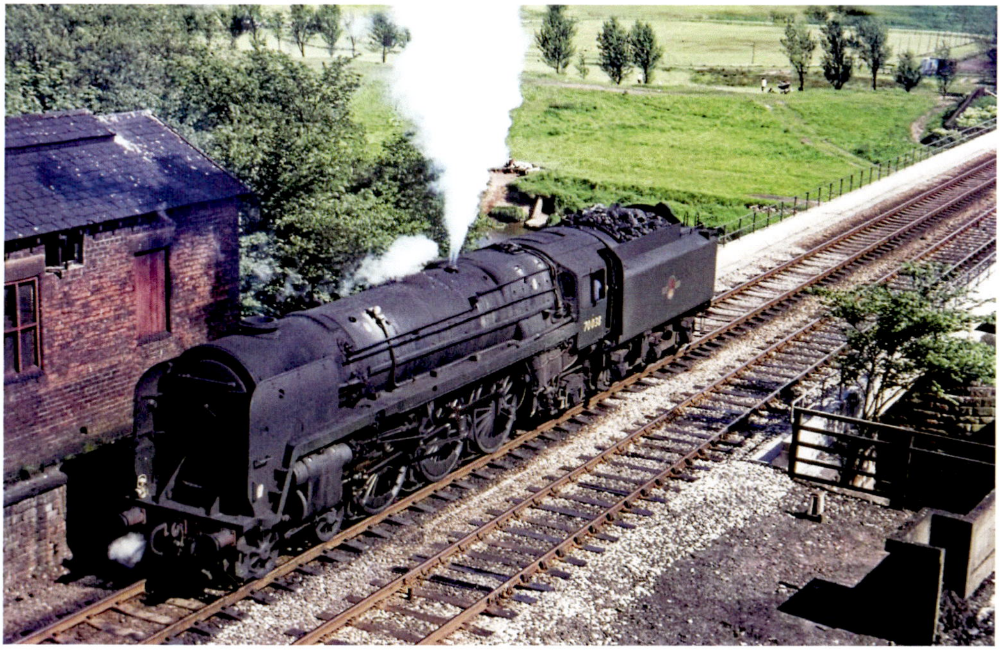

August 1965 BR Standard Britannia 7MT 4-6-2 70038 reverses over Eden Bridge, Carlisle.
BR Standard Britannia 7MT 4-6-2 70038 *Robin Hood* reverses across Eden Viaduct near Etterby Junction after leaving Kingmoor depot. It was heading for Citadel station to collect a northbound express to Glasgow. It had originally been allocated to the Great Eastern Region and then spent three years at Immingham MPD, before transfer to Carlisle Kingmoor in December 1963. It was withdrawn from service in August 1967 and scrapped in early 1968.

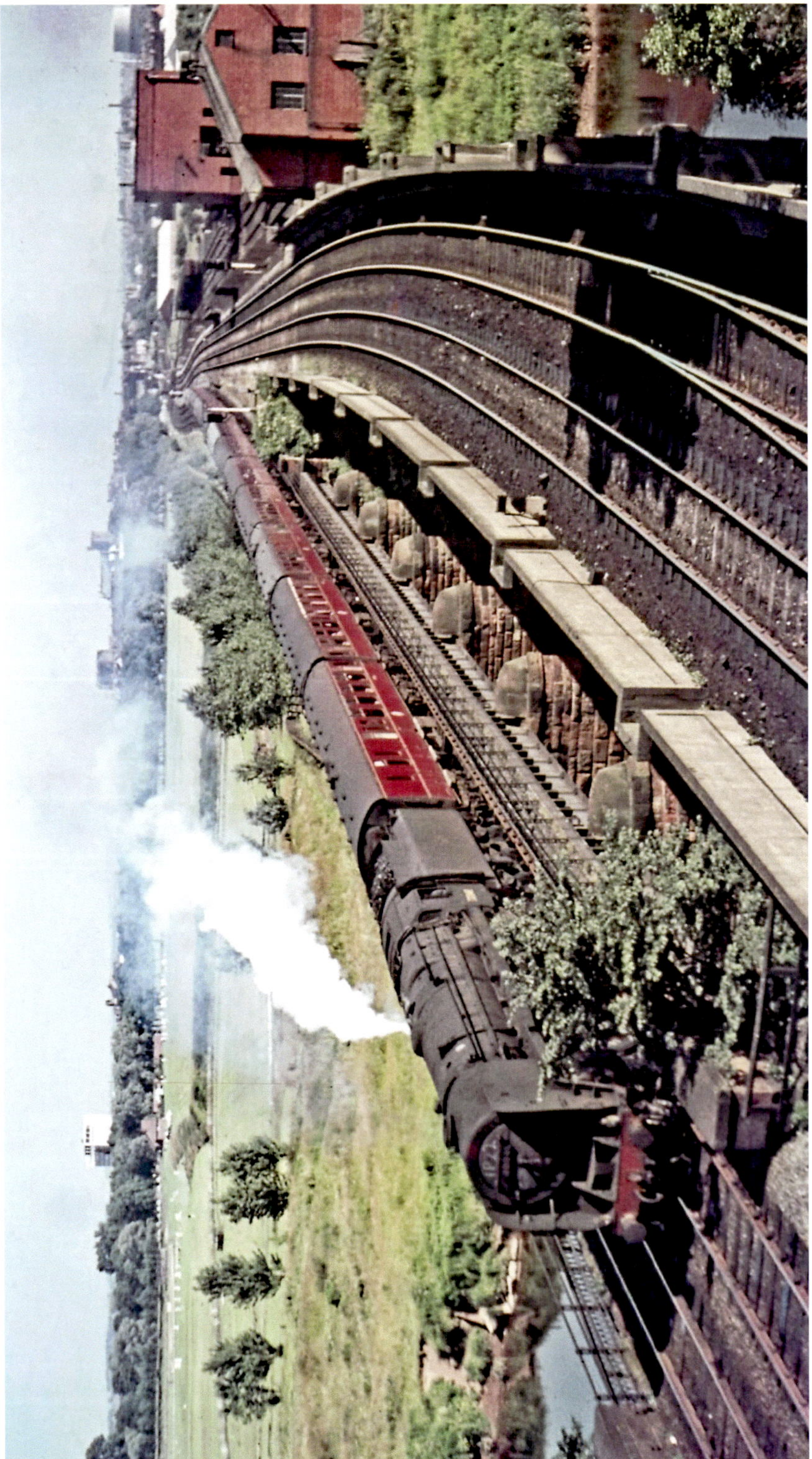

August 1965 BR Standard Britannia 7MT 4-6-2 70041 crosses Eden Bridge, Carlisle.

BR Standard Britannia 7MT 4-6-2 70041 *Sir John Moore* heads north over the River Eden with a summer Blackpool to Glasgow train which it attached to at Citadel station in the background. The train is seen heading over the original West Coast Main Line viaduct; the concrete viaduct in the photo was built in 1942 to cope with the heavy wartime freight movements. 70041 was to be withdrawn in April 1967 and was cut up by J. McWilliam of Shettleston in September 1967.

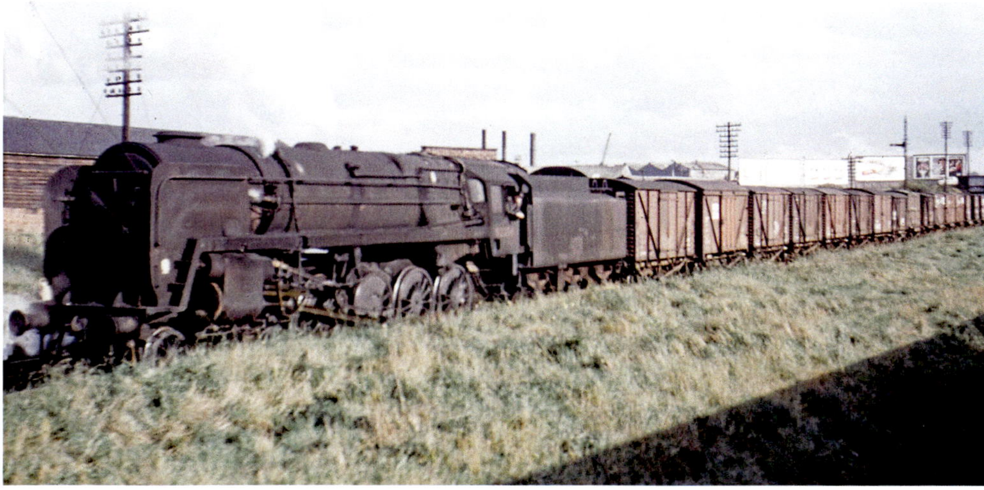

14 October 1967 BR 9F 2-10-0 92204 Carlisle with a northbound fitted freight.
BR 9F 2-10-0 92204 had been built at Swindon Works in April 1959. It had originally been allocated to Western Region depots, spending two years at Old Oak Common and three years at Tyseley. When it first entered traffic, it was sent to Bristol St Philips Marsh MPD, and while there was used on an eleven-coach trial train over the Somerset and Dorset line. The success of the trial led to 9Fs being regularly used on the line for the heavy summer passenger trains. In August 1966 it was transferred to Speke Junction MPD and was withdrawn from there in December 1967. It was cut up by T. W. Ward of Beighton in the spring of 1968.

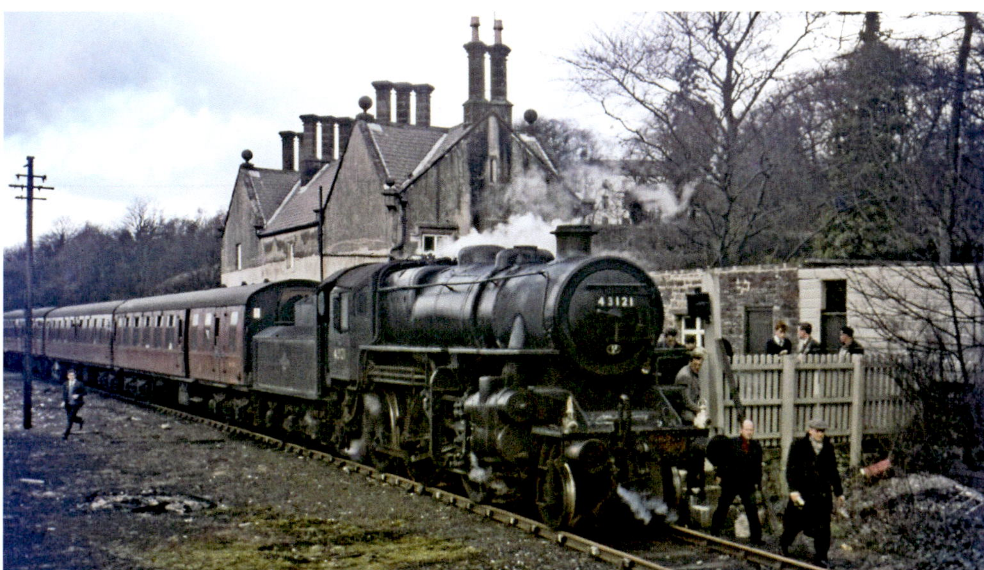

26 March 1967 LMS Ivatt 4MT 2-6-0 43121 at Alston.
LMS Ivatt 4MT 2-6-0 43121 had been built at Horwich and entered service in August 1951. For eleven years it was allocated to Cricklewood MPD before moving to Wellingborough and Heaton Mersey. Carlisle Kingmoor became its home in June 1965 until it was withdrawn from there in November 1967. It was seen at the head of the Stephenson Locomotive Society's Scottish Rambler which it had hauled from Carlisle to the branch line terminus of Alston.

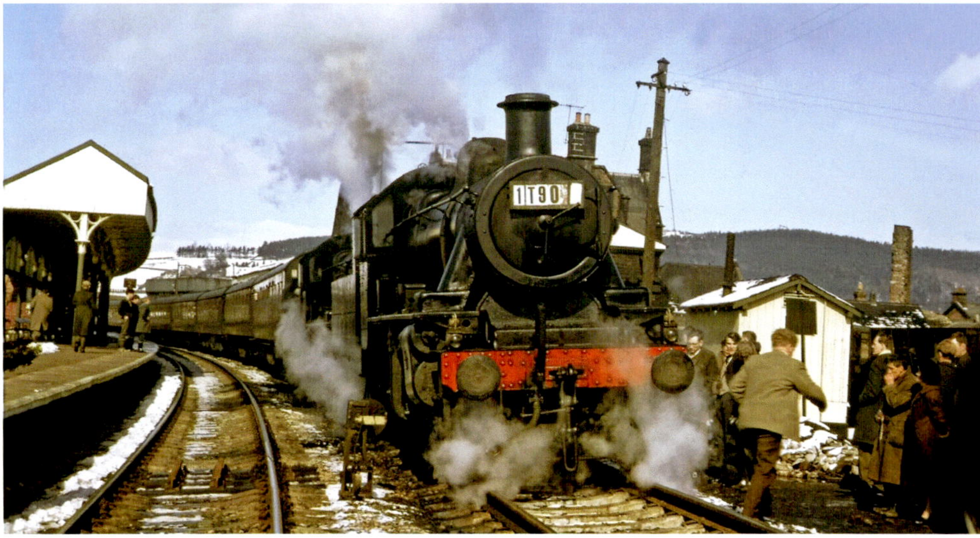

2 April 1966 LMS Ivatt 2MT 2-6-0s 46426 and 46458 at Penrith.
LMS Ivatt 2MT 2-6-0s 46426 and 46458 were both Carlisle Upperby-allocated locomotives which had been used before as suitable motive power for enthusiasts' specials on the lines around Penrith, Workington and Carlisle. On this occasion it was the Stephenson Locomotive Society Lakes & Fells Rail Tour which had been hauled from Manchester to Hellifield by 45596 and on to Penrith by preserved A3 4472 *Flying Scotsman*. The pair of Ivatt 2MTs took the special to Keswick, Cockermouth, Workington, Millom and Arnside. It was to be the last train hauled by steam to work west of Keswick as the section closed a few weeks later. At Arnside they handed over to 4472, which headed the train back to Hellifield and from there 45596 returned the enthusiasts back to Manchester.

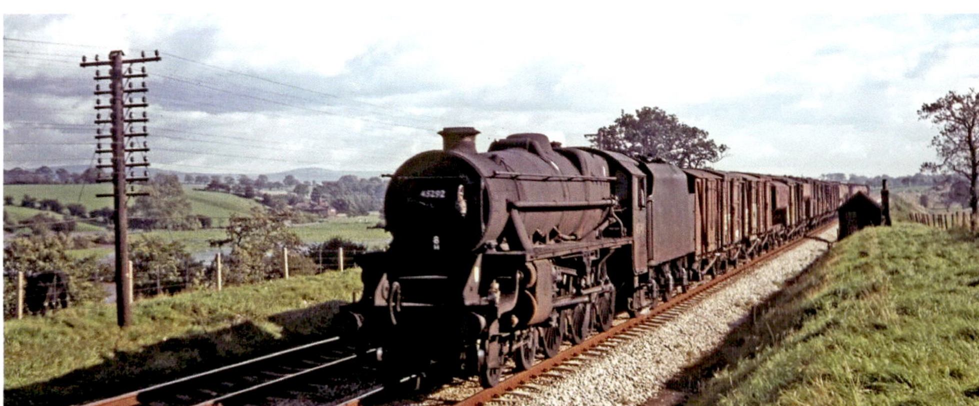

4 October 1967 LMS Black Five 4-6-0 45292 heads a fitted freight train south of Carlisle.
LMS Black Five 4-6-0 45292 was one of the batches of this class built by Armstrong Whitworth of Newcastle for LMS in 1936. In December 1936 it entered service in South Wales at Swansea, and later moved to North Wales where it spent three years at Llandudno Junction and three years at Holyhead. It was later based at Northampton, Bletchley, Aston, Stoke and Willesden. At the time of the photograph the hand-painted shed code denotes its last allocation which was at Birkenhead Mollington Street depot. It was a few weeks away from withdrawal at the beginning of November and was cut up by Cashmore's of Great Bridge in March 1968.

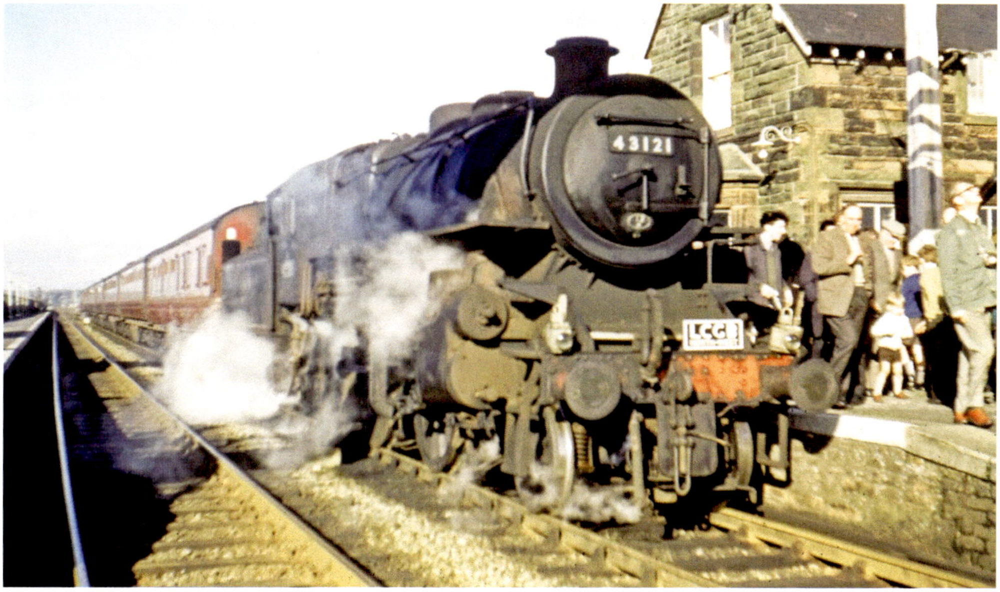

14 October 1967 LMS Ivatt 4MT 2-6-0 43121 with a LCGB special at Longtown Junction.
LMS Ivatt 4MT 2-6-0 43121 was used to haul the Locomotive Club of Great Britain special called the 'Castle to Carlisle' around the lines to the north of the city including part way along the Waverley route. The train had originated in Liverpool and was headed by BR 9F 2-10-0 92091 as far as Preston. Preserved GWR Castle 4-6-0 7029 *Clun Castle* took the train over the West Coast Main Line to Carlisle where 43121 took over. On return from Carlisle, 7029 worked over the Settle and Carlisle line to Hellifield where it handed over to 92091 for the run back to Liverpool.

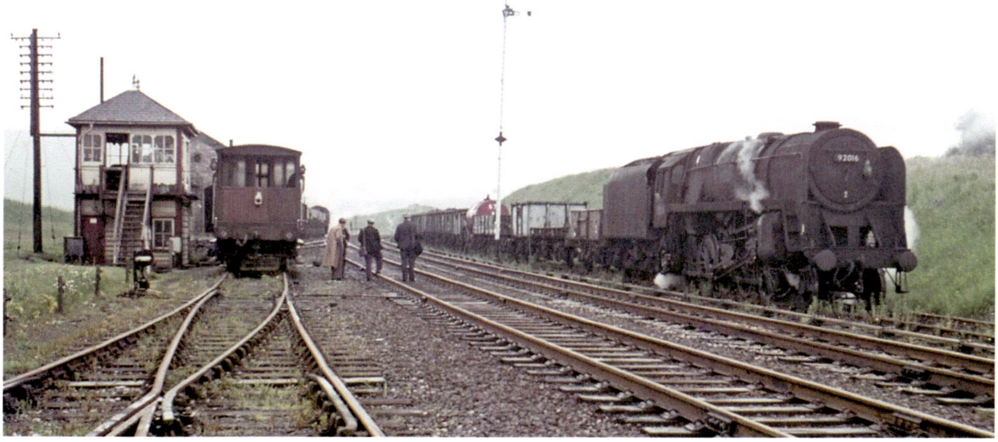

19 August 1967 BR 9F 2-10-0 92016 at Kirkby Stephen West.
BR 9F 2-10-0 92016 had been built at Crewe Works in October 1954 and spent three years allocated to Wellingborough MPD before it moved to the North West, making Newton Heath its home. In June 1967 it was reallocated to Carnforth and was seen in the loop at Kirkby Stephen with a mixed-goods train. Its crew was watching the departure of another 9F hauling a Long Meg to Widnes anhydrite working. This was the last summer of everyday steam working on the Settle and Carlisle line. 92016 was withdrawn in October and cut up in April 1968 by J. McWilliam of Shettleston. (Photograph L. Flint)

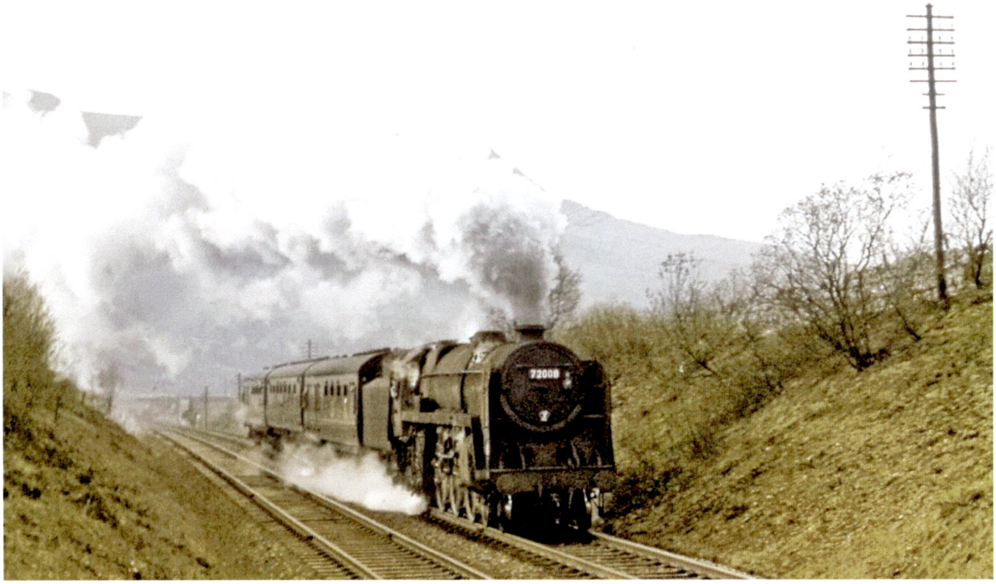

3 April 1965 BR 4-6-2 Clan 72008 climbing Ais Gill.
BR 4-6-2 Clan 72008 *Clan Macleod* climbing towards Ais Gill with its all stations stopping train bound for Bradford Foster Square. This was a regular afternoon duty for the survivors of the class at this time. 72008 sent a smokescreen high and wide almost blanking out Wild Boar Fell in the background. It was to soldier on for a further year, becoming the last of the class to be withdrawn.

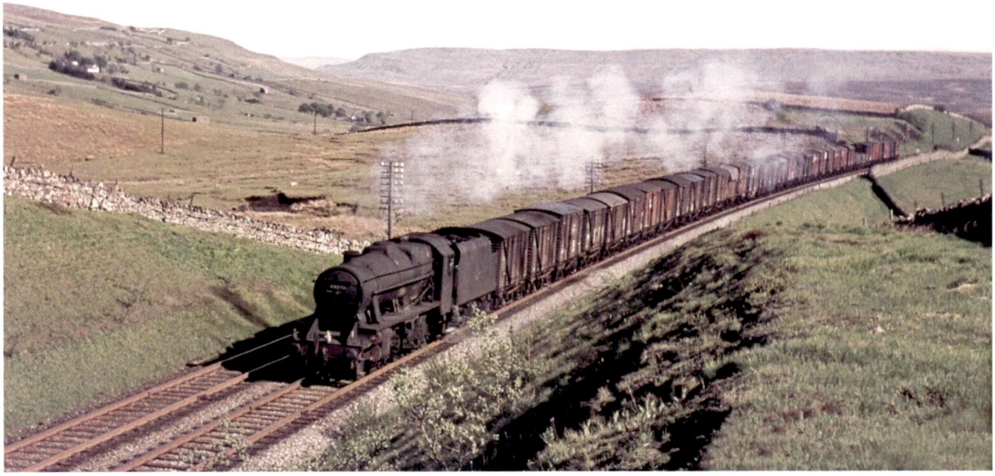

28 May 1966 LMS 8F 2-8-0 48276 near Ais Gill.
LMS 8F 2-8-0 48276 had been built by the North British Locomotive Company of Glasgow for the War Department and on completion had been loaned to the LMS. It was bought by the LMS in December 1943 and taken into British Railways stock in January 1948. It remained a West Yorkshire-based engine throughout its working life, which started at Stourton and included Mirfield, Low Moor, Wakefield and Royston. It was a Mirfield locomotive when seen climbing towards Ais Gill with a fitted van train. It was withdrawn from Royston in November 1967 and was cut up by Cashmore's of Great Bridge in April 1968. (Photograph L. Flint)

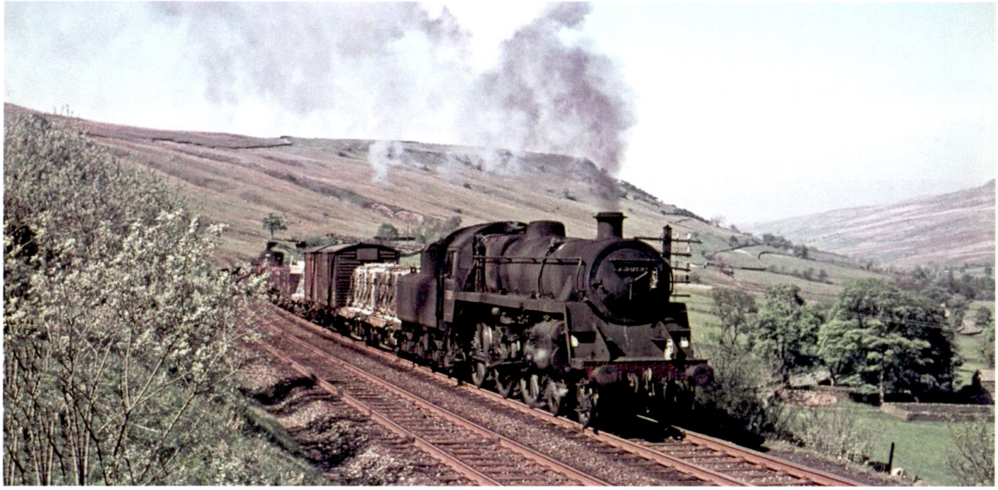

28 May 1966 BR Standard 4MT 4-6-0 75019 near Ais Gill.
BR Standard 4MT 4-6-0 75019 had been built at Swindon Works in March 1952 and entered service at Southport, which remained its home until January 1964. Its first reallocation was to Springs Branch, Wigan; however, just over one year later it was on its way to Skipton. It then spent six months on banking duties based at Tebay MPD, before its final move in January 1968 took it Carnforth from where it was withdrawn in August 1968, at the very end of steam on British Railways. Although there were thoughts about its private purchase for preservation, it was cut up by Campbell's of Airdrie. (Photograph L. Flint)

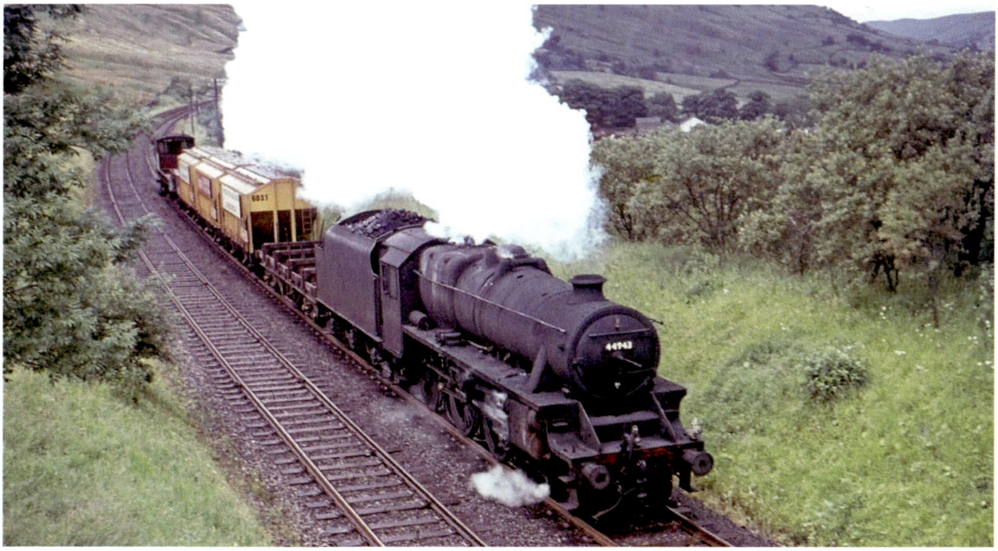

June 1967 LMS Black Five 4-6-0 44943 near Ais Gill.
LMS Black Five 4-6-0 44943 was a product of Horwich Works and entered traffic in December 1945. It had been based in West Yorkshire throughout its career and was a stalwart of Holbeck depot. It did have a three-year move to Farnley Junction MPD, but in December 1966 had returned to Holbeck from where it was withdrawn in September 1967. In February 1968 it was cut up by T. W. Ward at Killamarsh. It was seen making light work of the climb up Ais Gill with a short freight a couple of months before its demise. (Photograph L. Flint)

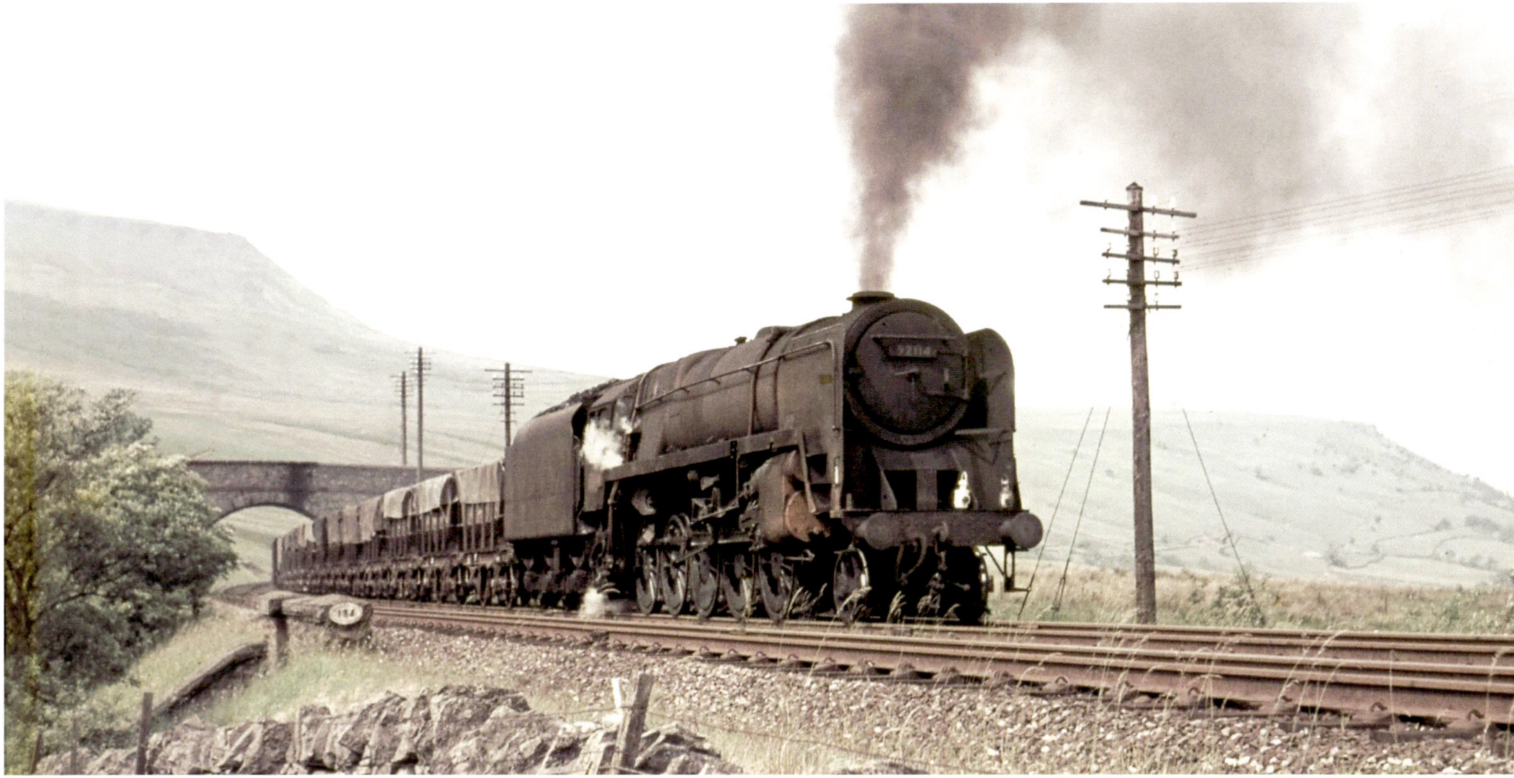

July 1967 BR 9F 2-10-0 92114 slogs up Ais Gill.
BR 9F 2-10-0 92114 had been built at Crewe Works and entered service at Westhouses MPD in December 1956. By June 1960 it had been transferred to Toton depot and two years later to Rowsley. After spending a year at Kirkby-in-Ashfield it was reallocated for the last time to Carlisle Kingmoor from where it was withdrawn in July 1967. It was seen making good progress towards the summit of Ais Gill at the head of a heavy Long Meg to Widnes anhydrite train. Although looking unkempt, it seemed to be in reasonable condition and was on top of the arduous task it was undertaking. It is a surprise that this may have been its last duty as it was withdrawn days later and stored at Kingmoor before being cut up by Campbell's of Airdrie in March 1968. (Photograph L. Flint)

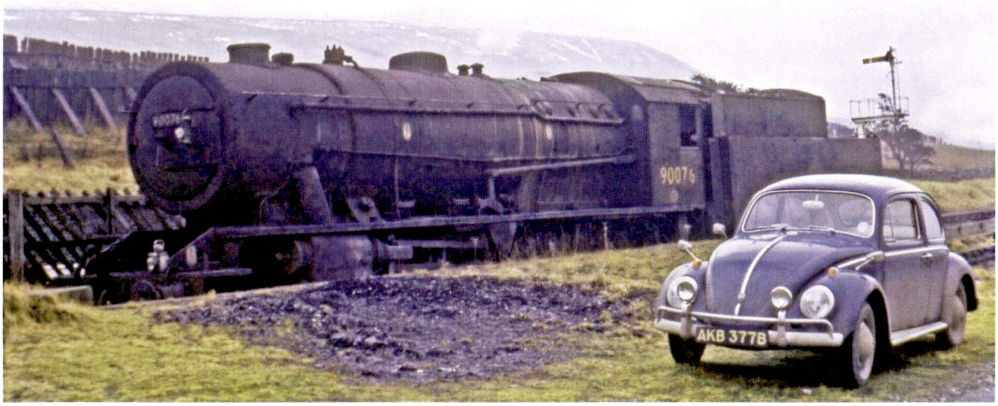

January 1967 WD 2-8-0 90076 at Dent station.
WD 2-8-0 90076 had been built by the North British Locomotive Company of Glasgow, for the War Department, as 7276, in August 1944, later to be numbered 77276. It was used by the USA Transportation Corps in north-west Europe but by November 1945 had returned to England and loaned to the LNER. In December 1946 it was bought by the LNER and at the creation of British Railways it was numbered 90076. It had always been allocated to the North Eastern Region at various depots including York, Darlington and Thornaby. In March 1959 it was transferred to Wakefield carrying the 'O' symbol on its cab side denoting its suitability for use on the tightly timed coal train turns. It had apparently failed while working over the Settle and Carlisle line and was left in a goods bay at Dent station. It was later recovered and went on to be transferred to Normanton in July 1967. It was withdrawn from West Hartlepool MPD a month later.

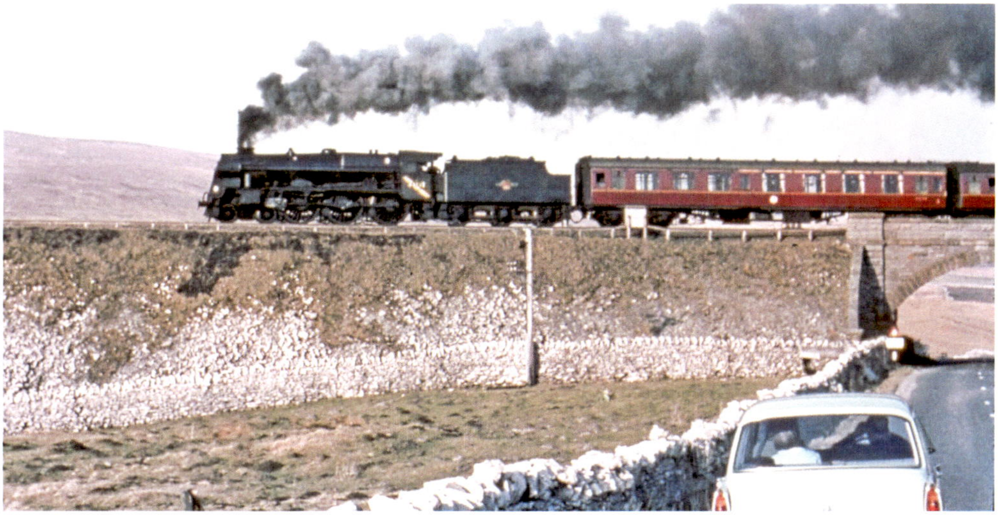

3 April 1965 LMS Royal Scot 4-6-0 46152 heads towards Ribblehead Viaduct.
LMS 4-6-0 Royal Scot 46152 *The King's Dragoon Guardsman* powers its train away from Ribblehead station and towards the viaduct while working the 15.40 Bradford Forster Square to Carlisle all stations stopper. Although the locomotive looks well presented, it was probably working its last train as it was withdrawn from service a few days later. The train consisted of the usual three coaches with the addition of seven other vehicles which included a recently outshopped Derby lightweight two-car diesel multiple unit. (Photograph L. Flint)

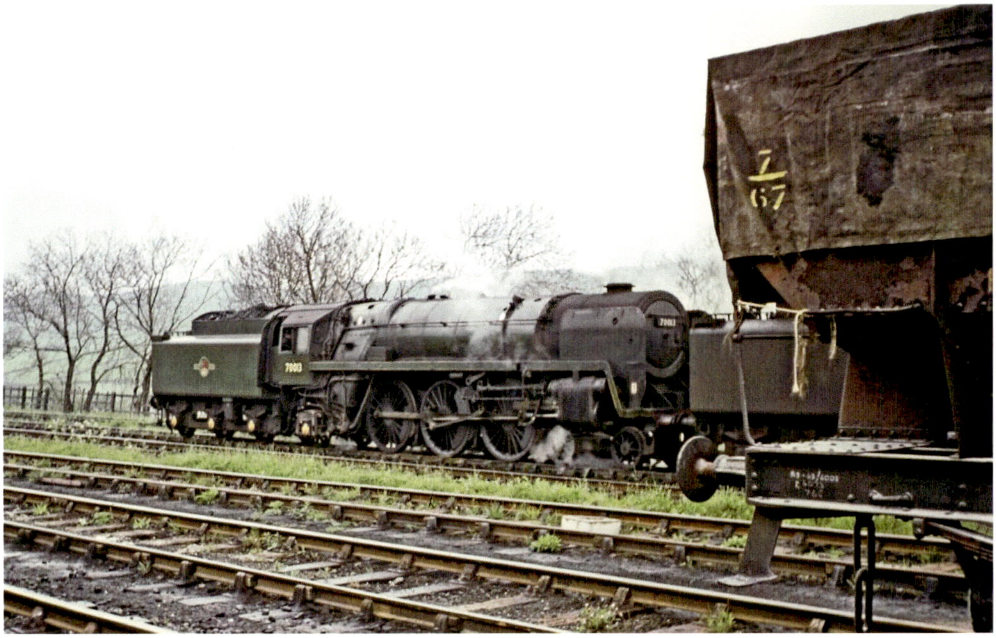

3 June 1967 BR Standard Britannia 7MT 4-6-2 70013 in the loop at Horton-in-Ribblesdale.
BR Standard Britannia 7MT 4-6-2s 70013 and 70023 had been dispatched from Carlisle Kingmoor to Hellifield to act as stand-by locomotives for the extra trains to be worked over the Settle and Carlisle line due to the diversion of traffic from the West Coast Main Line. The main line had been closed at Penrith for the installation of a new motorway bridge. Unfortunately, the leading engine, 70023, attached ahead of 70013, was derailed in the loop at Horton-in-Ribblesdale en route to Hellifield. The diversions which had been put in place were thrown into chaos as a result, because single-line working had to be implemented between Helwith Bridge and Blea Moor in order to rescue the marooned 'Britannias'.

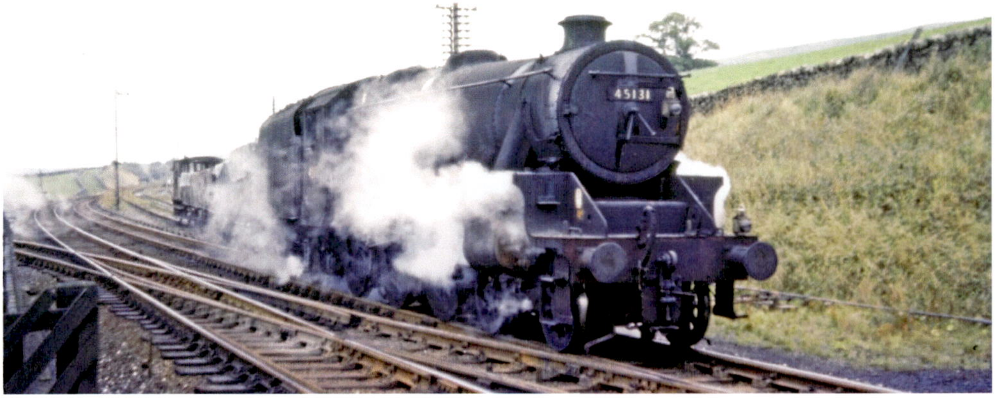

June 1967 LMS Black Five 4-6-0 45131 leaves the loop at Horton-in-Ribblesdale.
LMS Black Five 4-6-0 45131 had been built by Armstrong Whitworth of Newcastle for LMS in May 1935 and spent a long time allocated to both Crewe North and South depots. In October 1961 it was transferred to Edge Hill MPD and in October moved across to Speke from where it was withdrawn in April 1968. It was seen leaving the loop at Horton-in-Ribblesdale with a short train of minerals from Long Meg.

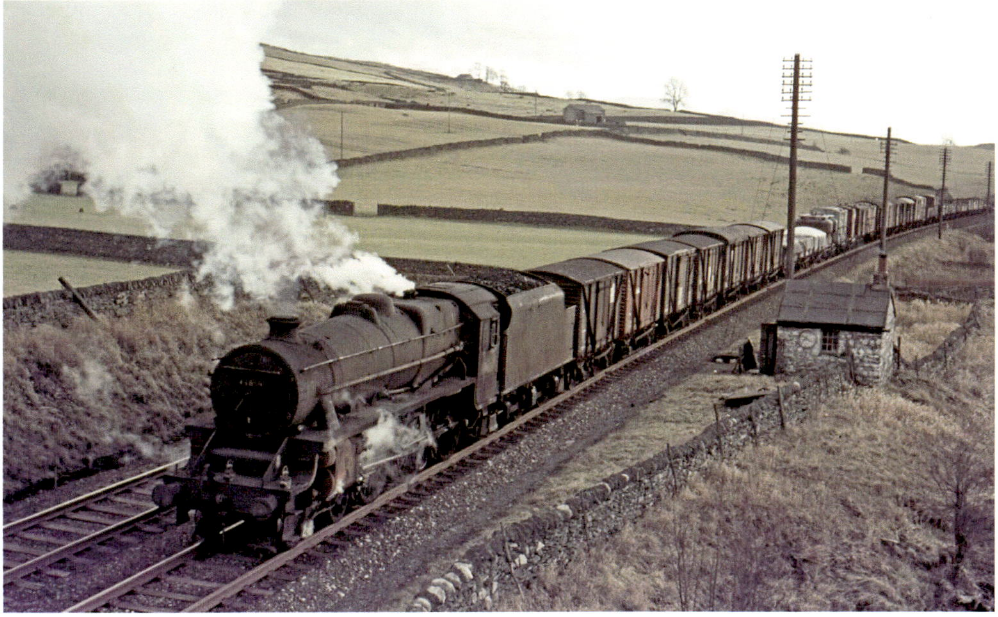

February 1966 LMS Black Five 4-6-0 44659 passes Helwith Bridge.
LMS Black Five 4-6-0 44659 had been built at Crewe Works in May 1949 for British Railways. It was based in the East Midlands for much of its working life at depots including Leicester Midland, Derby and Colwick. In April 1966 it was reallocated to Aintree MPD from where it was withdrawn in June 1967. It was seen working hard as it passed Helwith Bridge with a lengthy Stourton to Carlisle freight. (Photograph L. Flint)

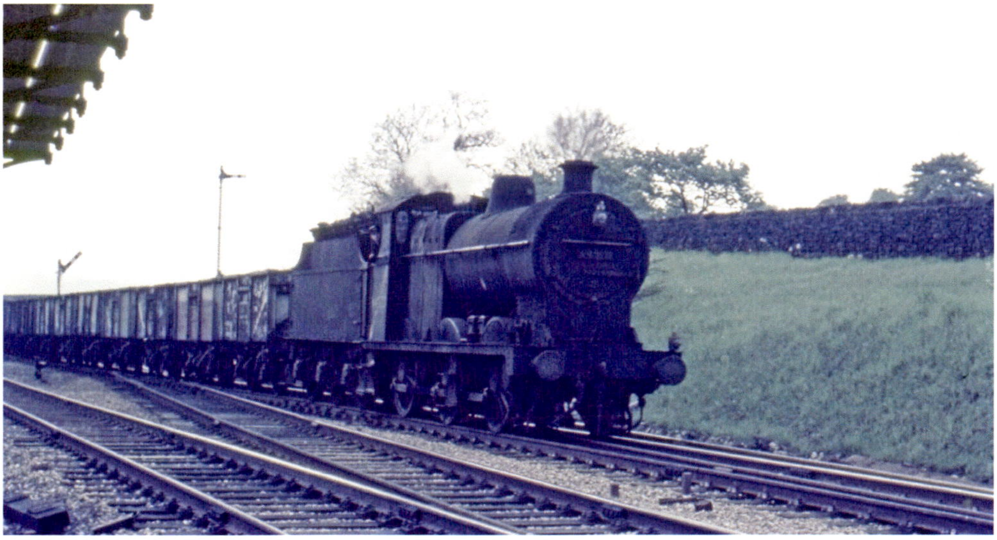

April 1965 LMS 4F 0-6-0 44570 runs through the junction at Settle station.
LMS 4F 0-6-0 44570 had been built at Crewe Works in August 1937 and was a long-term resident at depots in West Yorkshire – Normanton, Manningham and Skipton. It had been allocated to Stourton since October 1951 and was to remain there until withdrawn in October 1965. It was cut up late in 1965 by T. W. Ward at Beighton. (Photograph L. Flint)

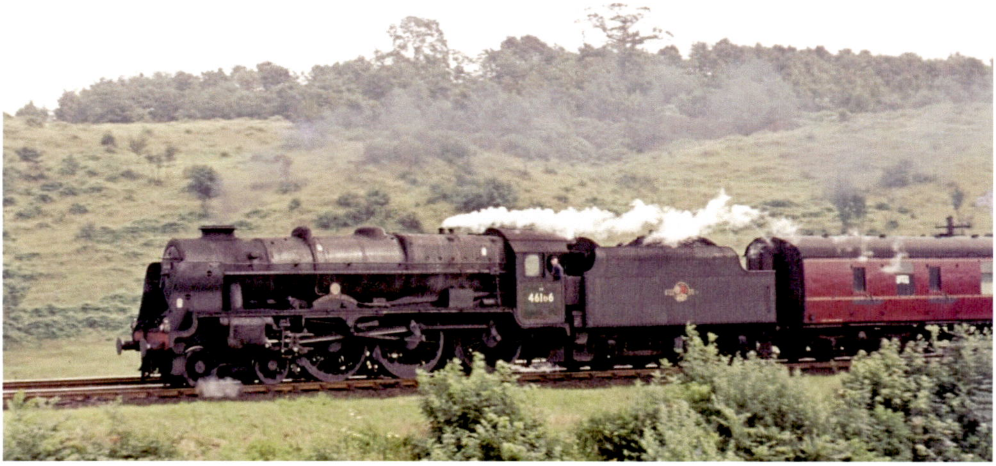

June 1964 LMS Royal Scot 4-6-0 46166 hurries down the bank near Oxenholme.
LMS Royal Scot 4-6-0 46166 *London Rifle Brigade* was built at Derby in October 1930 and rebuilt in the form seen here in January 1945. It had spent most of its career at Crewe North depot after being rebuilt until 1960 when it was transferred to Longsight MPD and then in 1963 to Carlisle Upperby. It was seen near Oxenhope with the crew taking advantage of the downhill gradient to recover from the climb up Shap. The locomotive was still carrying its nameplates but had not yet received the diagonal yellow warning stripe on the cab. It did wear a stripe from August 1964 but was to be withdrawn in September of the same year and cut up by West of Scotland Shipbreaking of Troon in December.

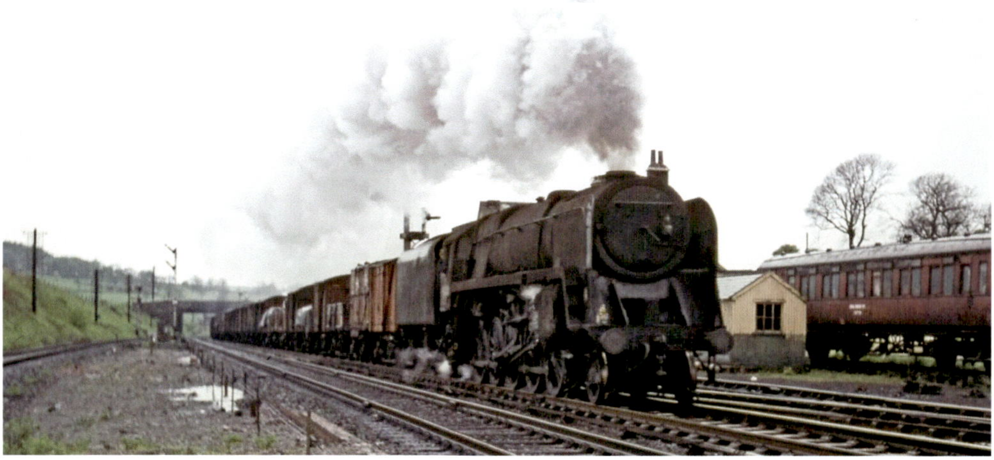

July 1966 BR 9F 2-10-0 thunders through Grayrigg.
An unidentified BR 9F 2-10-0, which had already lost its smokebox numberplate, thunders through the site of Grayrigg station with a long, heavy fitted freight. The station had closed in February 1954 and was at the summit of a continuous, 20-mile climb northwards from Carnforth. The 9F appears to be in fine health and ready to tackle Shap, the next big challenge, a few miles ahead, on its journey to Carlisle. The exhaust from the banking engine is just visible as it prepares to drop off the train.

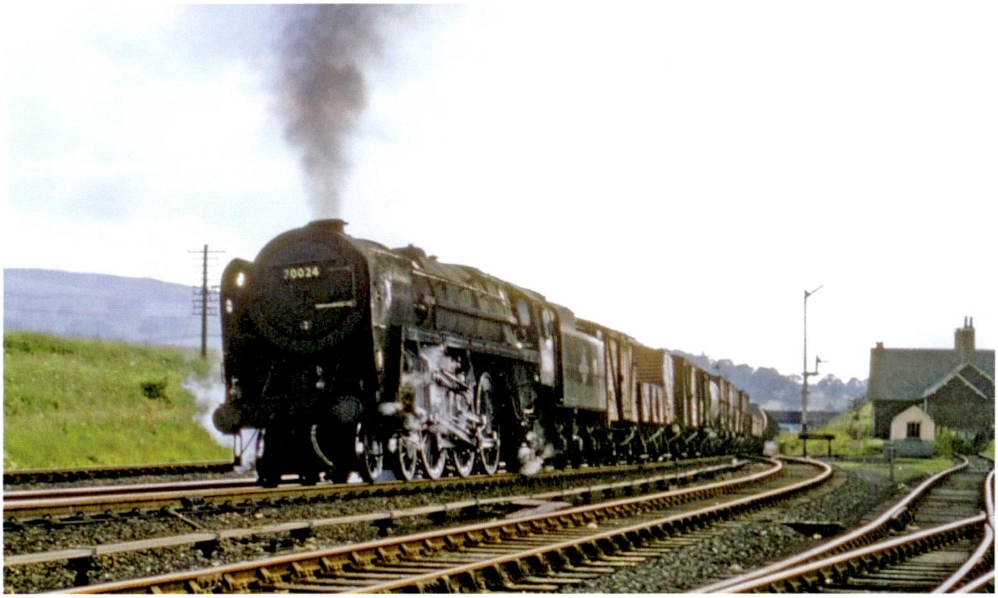

14 July 1966 BR Standard Britannia 7MT 4-6-2 70024 storms through Grayrigg northbound.
BR Standard Britannia 7MT 4-6-2 70024 *Vulcan* was built at Crewe and entered service at Plymouth Laira in October 1951. It was moved to Cardiff Canton in December 1956. Between September 1961 and its arrival at Carlisle in August 1966 it had a very nomadic career. It had been allocated to Aston, Rugby, Willesden, Crewe North, Holyhead, Llandudno Junction and Crewe South in that five-year period. After the closure of Carlisle Upperby in December 1966 it moved to Kingmoor and stayed until the depot closed in December 1967. It was then withdrawn, stored at Speke Junction MPD and then cut up by T. W. Ward at Killamarsh in June 1968.

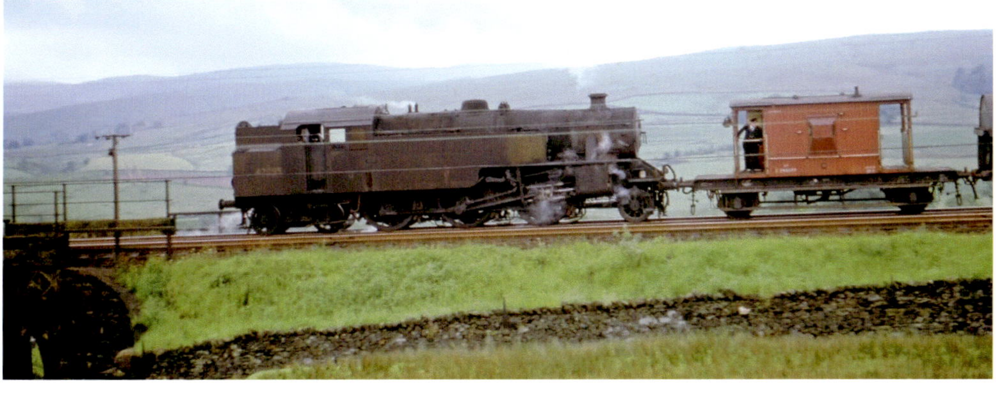

13 June 1964 LMS Fairburn 4MT 2-6-4T 42095 banking on Shap.
LMS Fairburn 4MT 2-6-4T 42095 was one of the batches of the class which had been constructed at Brighton Works in 1951. It had been allocated to Ashford until January 1958 when it moved to Dover Marine MPD. In December 1959 the locomotive was part of an inter-regional exchange with Fairburn 2-6-4Ts going to the London Midland Region and BR Standard 2-6-4Ts going to the Southern. Watford became the home of 42059 but it was reallocated to Carlisle Canal in July 1961. With the closure of Carlisle Canal in June 1963 it found a home at Tebay MPD to become a banking engine up Shap bank. It was withdrawn from these duties in June 1966 and was scrapped by Cashmore's of Great Bridge in October 1966.

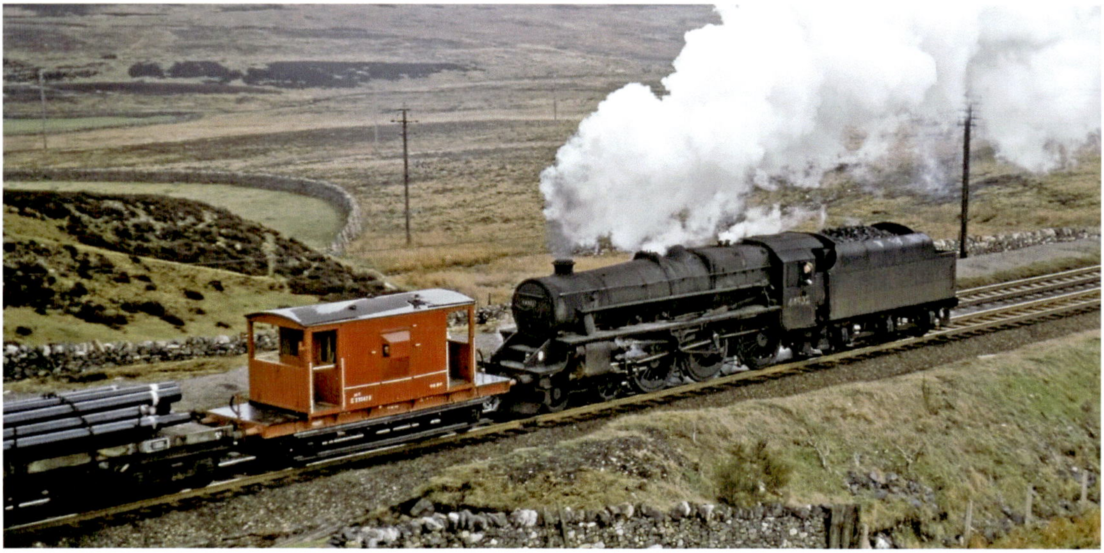

15 April 1965 LMS Black Five 4-6-0 44902 banking a goods train on Shap.
LMS Black Five 4-6-0 44902 was built at Crewe in October 1945, and unlike most other locomotives featured in this book, it was only ever allocated to one depot throughout its career. It gave twenty-two years' unbroken service to Carlisle Kingmoor, from where it was withdrawn in October 1967. It was cut up by J. McWilliam in Shettleston in March 1968. It was seen banking on Shap, no doubt drafted in to cover a temporary motive power shortage.

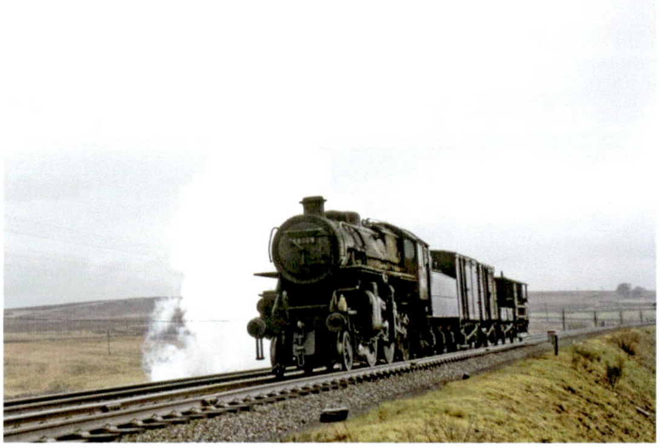

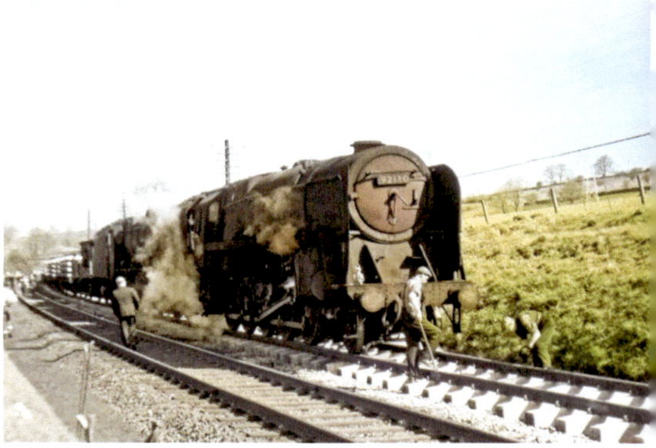

***Above left*: 15 April 1965 LMS Ivatt 4MT 2-6-0 43009 works up Shap with a short goods train.**
LMS Ivatt 4MT 2-6-0 43009 had been built at Horwich Works and entered service at Workington in March 1948. It made only one move in its career and that was to Tebay MPD in September 1960, where it was a regular performer banking trains up Shap until its withdrawal in September 1966. It was scrapped by J. McWilliam of Shettleston in the summer of 1967. It was seen making the ascent up Shap look easy with a lightweight pick-up goods train.

***Above right*: 1 May 1966 BR 9F 2-10-0 92130 helps with track relaying on Shap.**
BR 9F 2-10-0 92130 was built at Crewe Works in April 1957 and began work from Toton depot. Its only move came in May 1964 to Carlisle Kingmoor. It was seen on Shap together with an unidentified LMS 8F 2-8-0 at the head of a track-laying train. 92130 was actually withdrawn in May 1966, so this may well have been its last duty. It was cut up by Motherwell Machinery and Scrap of Wishaw in September 1966.

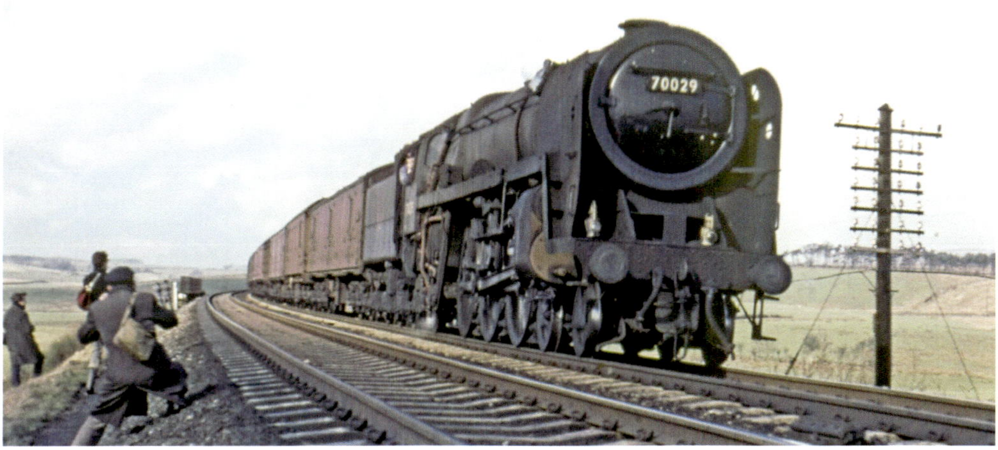

26 February 1967 BR Standard Britannia 7MT 4-6-2 70029 coasts downhill on Shap.
1967 BR Standard Britannia 7MT 4-6-2 70029 *Shooting Star* was built at Crewe Works and allocated to Cardiff Canton in November 1952. In September 1961 it was reallocated to Aston and four years later it moved to Carlisle Upperby. With the closure of Upperby depot it was transferred across the city to Kingmoor from where it was withdrawn in October. The idea of photographers being so close to the main line nowadays seems totally inconceivable yet in the late 1960s it became an everyday occurrence.

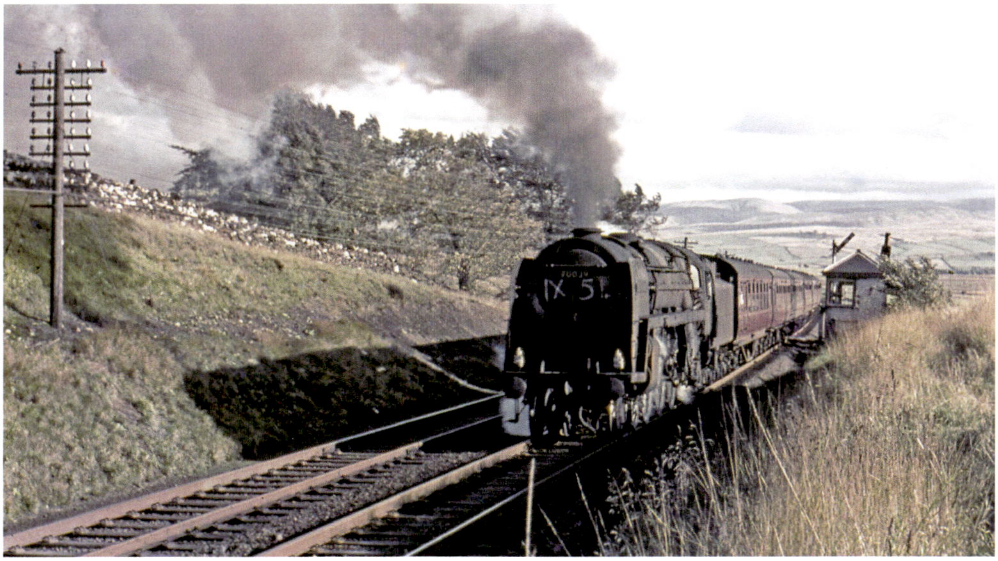

28 September 1964 BR Standard Britannia 7MT 4-6-2 70039 climbs past Scout Green.
BR Standard Britannia 7MT 4-6-2 70039 *Sir Christopher Wren* was built at Crewe Works in February 1953 and was allocated to the Great Eastern Region at Norwich Thorpe and then Stratford. After the dieselisation of services from Norwich to London, it was reallocated to Immingham MPD in December 1960 for three years, before being moved to Carlisle Upperby for a couple of months. It then moved on to Kingmoor and was there until its withdrawal in September 1967. It was seen as it thundered by the signal box at Scout Green, climbing northwards on the 1 in 75 gradients of Shap bank. 70039 still carried its nameplates, and climbed without the assistance of a banking engine. (Photograph L. Flint)

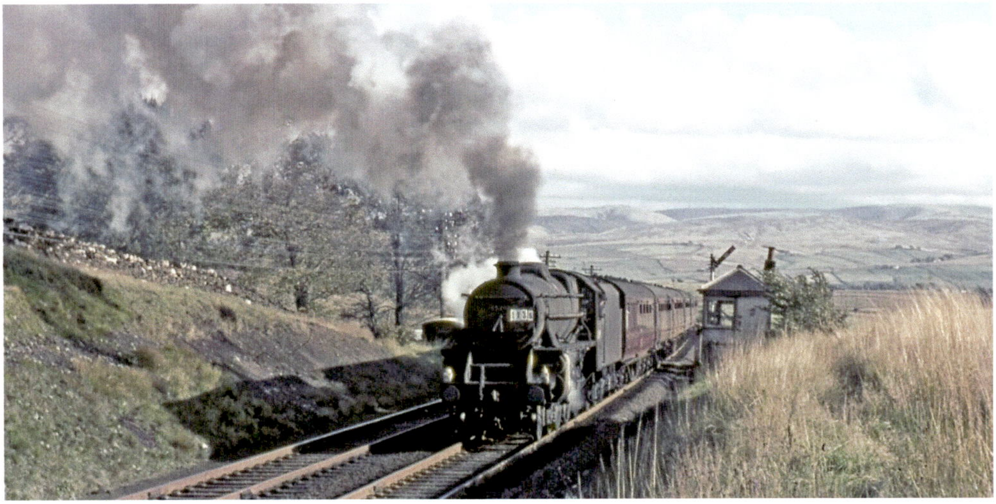

28 September 1964 LMS Jubilee 4-6-0 45703 climbs past Scout Green.
LMS Jubilee 4-6-0 45703 *Thunderer* seemed to be aptly named as it heads by Scout Green signal box on its ascent of Shap with a long, late summer extra. It was ably assisted at the rear by LMS Fairburn 4MT 2-6-4T 42210 which had been attached at Tebay, at the foot of the northern climb. 45703 had been built at Crewe Works in May 1936 and had many home depots in its career. It was reallocated often, twice a year sometimes, and reappeared on the same depot's books several times. Bushbury was its home four times, with the longest stay of three years on one occasion. Crewe North was another instance where it returned three times with the longest stint of four years. In its last days it was allocated to Carnforth and was eventually withdrawn two months after this photograph from Warrington Dallam. (Photograph L. Flint)

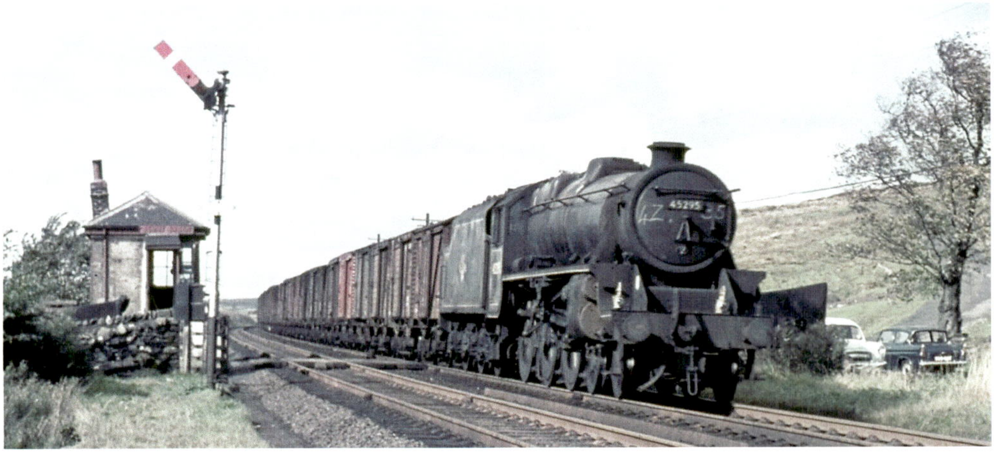

28 September 1964 LMS Black Five 4-6-0 45295 heads downhill at Scout Green.
LMS Black Five 4-6-0 45295 rushing past the signal box at Scout Green with its safety valves lifting at the head of a lengthy fitted freight from Kingmoor yards. 45295 had been built by Armstrong Whitworth for the LMS in December 1937. From the formation of British Railways in January 1948 it had been allocated to Carlisle Upperby. Its only move, apart from a three-month loan to Edge Hill, came in June 1963 across the city to Kingmoor. In December 1967 with the closure of Carlisle Kingmoor the locomotive was withdrawn and cut up in the spring of 1968 by T. W. Ward of Beighton.

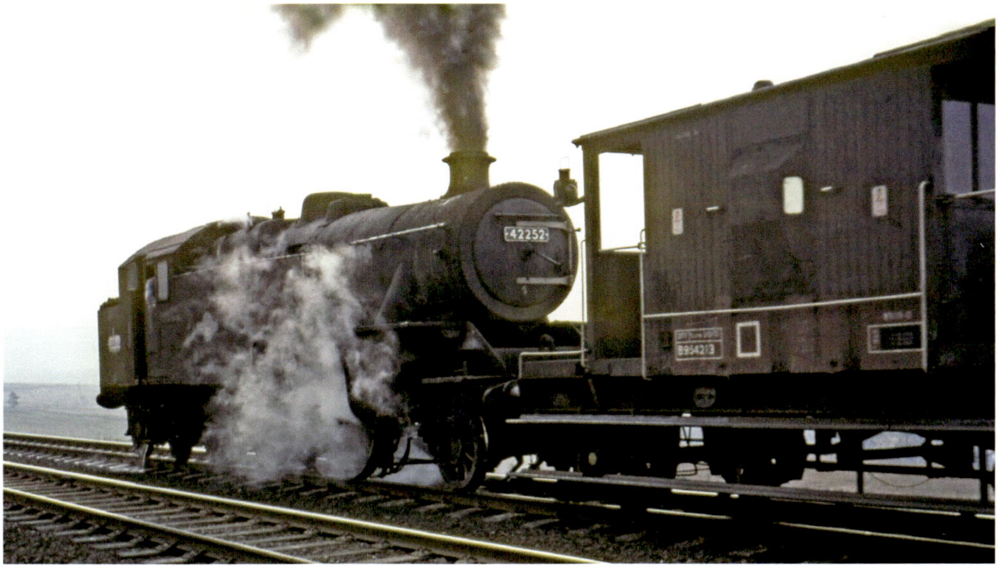

1 April 1967 LMS Fairburn 4MT 2-6-4T 42252 banking on Shap at Scout Green.
LMS Fairburn 4MT 2-6-4T 42252 was seen assisting a heavy northbound freight on the climb up Shap bank. The locomotive had been built at Derby Works in November 1946. It had spent fourteen years hauling commuter trains around the capital, seven years at Plaistow and then on to Neasden. Modernisation had made it redundant from these duties and it was transferred to Woodford Halse and then several other locations before finding its way to Tebay MPD in December 1966. It was to be replaced on its banking duties by BR Standard 4 4-6-0s. In June 1967 it was reallocated to West Yorkshire to Low Moor and Normanton from where it was withdrawn in September 1967.

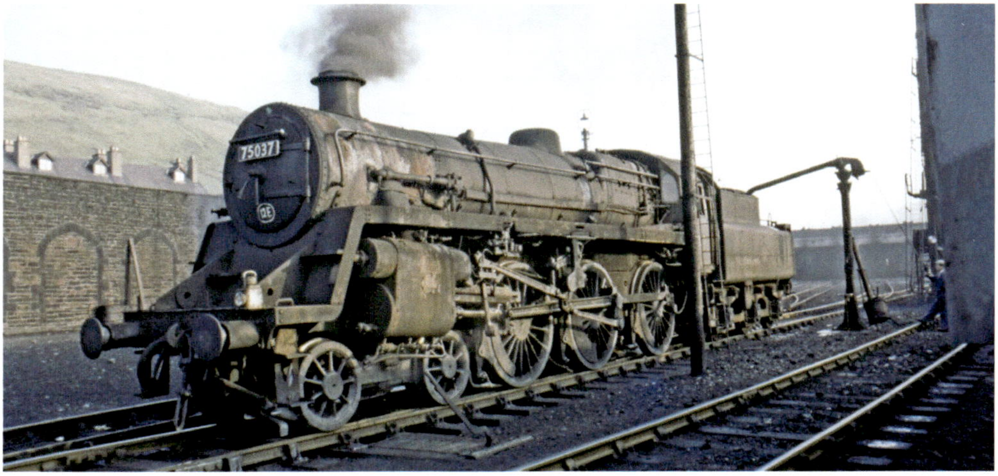

13 September 1967 BR Standard 4MT 4-6-0 75037 takes Water at Tebay MPD.
BR Standard 4MT 4-6-0 75037 had been built at Swindon in August 1953 and had spent nine years at Bletchley MPD before reallocation to Stoke for a four-year stay. In May 1967 it moved to Tebay depot to take up banking duties on Shap bank. With the cessation of steam banking in December 1967 it was withdrawn and after a period in store was finally cut up by Arnott Young of Carmyle in July 1968.

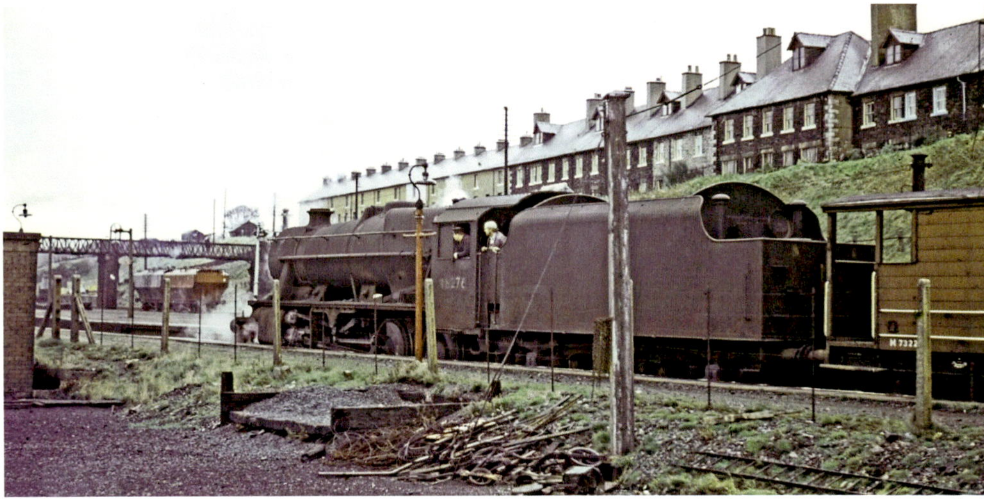

20 October 1967 LMS 8F 2-8-0 48276 restarts from Tebay station.
LMS 8F 2-8-0 48276 had been built for the War Department in 1942 but sold to the LMS in 1943. It had always been allocated to sheds in West Yorkshire and spent twelve years at Stourton MPD and six years at Mirfield. In July 1967 it had been reallocated to Royston depot and was seen at Tebay station at the head of a heavy prefabricated track panel train. BR Standard 4MT 4-6-0 75032 had been attached at the rear and the cavalcade had just restarted ready to tackle the gruelling climb up Shap. The locomotive was withdrawn on 4 November 1967, with the closure of Royston depot, and after heading the last freight working from Royston the day before. (Photograph L. Flint)

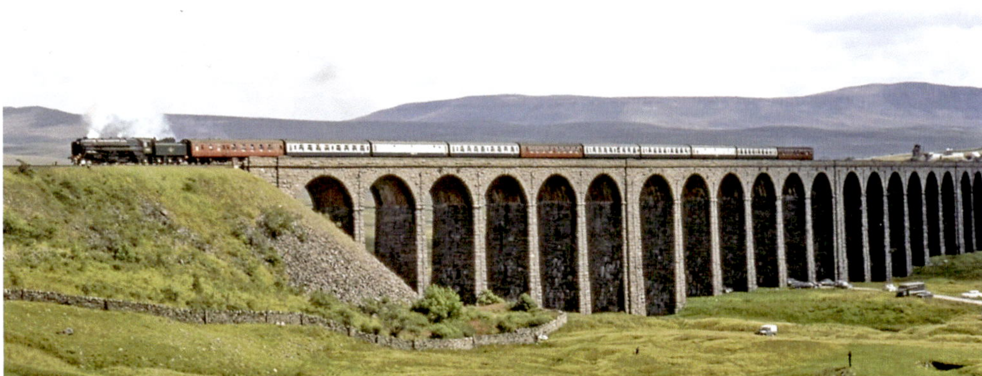

11 August 1968 BR Standard Britannia 7MT 4-6-2 70013 steams across Ribblehead Viaduct.
BR Standard Britannia 7MT 4-6-2 70013 *Oliver Cromwell* crossed Ribblehead Viaduct at around 14.00 at the head of the '15 Guinea Special'. The train had brought the curtain down on the use of steam power on British Railways. It had originated at Liverpool Lime Street and was hauled to Manchester Victoria by LMS Black Five 4-6-0 45110. The second leg of the journey was entrusted to 70013 and went from Manchester to Carlisle, via Hellifield, and over the Settle and Carlisle Railway. Later the same day 70013 was seen crossing back over Ribblehead light-engine, as it made its way to Bressingham and on to preservation. The special also retraced its journey across Ribblehead to Manchester hauled by LMS Black Five 4-6-0s 44781 and 44871.